T0419384

CYTOCHROME P450 ENZYMES

BIOCHEMISTRY, PHARMACOLOGY AND HEALTH IMPLICATIONS

BIOCHEMISTRY RESEARCH TRENDS

Additional books in this series can be found on Nova's website under the Series tab.

Additional e-books in this series can be found on Nova's website under the e-book tab.

CYTOCHROME P450 ENZYMES

BIOCHEMISTRY, PHARMACOLOGY AND HEALTH IMPLICATIONS

JIAN WU, M.D., PH.D.
EDITOR

nova publishers

New York

For permission to use material from this book please contact us:
Telephone 631-231-7269; Fax 631-231-8175
Web Site: http://www.novapublishers.com

NOTICE TO THE READER

Library of Congress Cataloging-in-Publication Data

ISBN: 978-1-61942-209-4

Library of Congress Control Number: 2014946531

Published by Nova Science Publishers, Inc. † New York

To my wife: Yao and Daughter: Lucy.

CONTENTS

PREFACE

Cytochrome P450 (CYP) is a super family of phase I enzyme in the biotransformation of xenobiotics and medications. Most medications undergo deactivation by CYP, and then are eliminated through either bile or kidneys from the body. CYP isozymes play a crucial role in drug interactions that may result in enhanced toxicity, reduced efficacy or onset of adverse reactions. On the other hand, many agents affecting CYP expression and activity may alter metabolic rate of different medications co-administrated. Therefore, the molecular basis, regulation by inducers or inhibitors, and pharmacologic reaction of specific CYP isozymes are the key issues of biochemical mechanisms, pharmaceutical development and safe use of various medications. This book is to meet the needs from basic molecular biochemists, pharmacologists, pharmacists, medical students, clinical practitioners and scientists, as well as broad readers who wish to understand how an herbal extract, medication or natural supplement is metabolized or transformed in the liver or other sites for deactivation and elimination. Special focuses are paid to herbal extracts and medications in the treatment of neuro-psychiatric or cardiovascular disorders, diabetes and viral hepatitis. Detailed dissection of drug interactions in a particular field intends to provide rationales for useful guidance of safe drug use in daily practice. The contributing authors are basic scientists, pharmacists, pharmacologists and on-service physicians in cardiovascular, neuro-psychiatric, gastroenterologic and hepatologic fields from Europe (Germany, France, Portugal), Australia, the US and China. Thus, the book is the collection of master pieces by well-known experts from various regions of the world, and represents the current understanding of CYP enzyme reaction and a contemporary coverage of possible drug interactions in involved fields. The featured chapters are scientific elucidation of basic biochemistry, pharmacology and clinical investigations in the interest of drug metabolism, interaction and safe use guidance in the single focus of this microsomal enzyme with multi-facet metabolic function.

Jian Wu, M.D., Ph.D.
Fudan University Shanghai Medical College,
Shanghai, China

In: Cytochrome P450 Enzymes
Editor: Jian Wu

ISBN: 978-1-61942-209-4
© 2014 Nova Science Publishers, Inc.

Chapter 1

CYTOCHROME P450:
ELECTRON TRANSFER AND SENSORS

Frieder W. Scheller[1,2,*], Aysu Yarmana[2]
and Ulla Wollenberger[2]

[1]Fraunhofer Institute for Biomedical Engineering IBMT,
Potsdam, Germany
[2]Institute of Biochemistry and Biology, University of Potsdam,
Potsdam, Germany

ABSTRACT

Electrochemical methods allow to characterize the reaction mechanism of the cytochrome P450 (CYP) enzymes by observing the electron transfer in real time. According to the number of publications on protein electrochemistry, CYP enzymes have the third position after glucose oxidase and cytochrome *c*. Sensors for drug monitoring using different CYP enzymes are appropriate tools since CYP enzymes act on more than 90 percent of all drugs currently on the market. For sensor development the efficiency of coupling the biocatalytic systems with the electrode is the crucial step. Furthermore for screening drugs, and for assessment of toxicity and prediction of drug clearance, the distribution of CYP isoenzymes and polymorphic enzymes would be of high clinical relevance.

---

[*] Corresponding Author address: Email: fschell@uni-potsdam.de.

Keywords: Cytochrome   P450,   biosensors,   personalised   medicine, bioelectrocatalysis, drug monitoring

INTRODUCTION

The catalytic cycle of CYP enzymes requires electron supply to the heme iron in the presence of oxygen. The majority of CYP enzymes accepts exclusively NAD(P)H as the source of electrons. However, for some CYPs an alternative pathway has been identified with peroxides as the source of activated oxygen (Denisov et al., 2005). This so called "peroxide shunt" pathway offers significant advantages for the preparative but also analytical applications. "Single oxygen donors" instead of molecular oxygen, such as sodium hypochlorite, iodosyl benzene, hydrogen peroxide, and cumene hydroperoxide, are used in order to eliminate the reductive activation of molecular oxygen and to circumvent the problems of direct electron exchange between the electrode and the protein in the presence of the natural cosubstrate oxygen (Rabe et al., 2008). The drawback of this reaction pathway results from the faster inactivation of the enzyme by the peroxide. To circumvent the destruction of the protein, continuous electrochemical or photochemical generation of peroxide has been applied for preparative substrate conversion (Girhard et al., 2013).

STATE OF THE ART OF CYP BASED SENSORS

CYP based sensors used biological systems of different complexity including animals (Hung et al., 2012), whole cells (e.g. hepatocytes) and microsomes but also the isolated native and mutant CYPs, chimeric proteins and alternative enzymes like peroxygenase (Table 1).

Table 1. Cytochrome P450 Based Sensors

Ref.	CYP species	Technique	Electrode modification	Substrates tested – catalysis	Comments
Renneberg et al., 1978	CYP liver microsomes	Amperometry	Carbon paste electrode	Aniline	First CYP sensor

Ref.	CYP species	Technique	Electrode modification	Substrates tested – catalysis	Comments
Iwuoha et al., 1998	CYP101	CV, amperometry	DDAB-BSA-gluteraldehyde - GCE	Catalysis with camphor, adamantanone and fenchone	$K_M' = 1.41$-3.9 mM. Use of $Co(Sep)^{3+}$, DET
Shumyantseva et al., 2001	RfCYP1A2 RfCYP2B4 RfCYPscc	CV, amperometry, spectrophotometry	Screen printed thick film Rh-graphite electrode	Aminopyrine, aniline. 7-ethoxyreso-rufin, 7-pentosyreso-rufin	Covalently attached riboflavin (Rf) to improve catalysis. Catalytic current at -500 mV
Krishnan et al., 2011	CYP1A2 CYP 2E1	CV, HPLC	LbL of CYP and reductase with PDDA on PGE	Conversion of 4-Methylnitros-amino)-1-(3pyridyl)-1-butanone	Electron supply via reductase
Panicco et al., 2011	CYP2D6 variants, CYP2C9 variants	CV, amperometry, HPLC, FTIR	Electrochemical array, polycrystalline Au electrodes	Bufaralol, warfarin	Fusion proteins with flavodoxin
Carrara et al., 2011	CYP 3A4,2B4,2C9	CV, amperometry	MWCNT on graphite electrode	Catalysis with benzphetamin, cyclophosphamide	Enhanced sensitivity by MWCNT
Wu et al., 2011a	CYP1A1	CV	EPG	Benzo(a) pyrene	Increase of oxygen current
Iwuoha et al., 2004	CYP2D6	CV, amperometry	Polyaniline dopedGCE	Catalysis with fluoxetine	$K_M' = 3.7$ µmol/L. $E°'$ shifted anodically in the presence of substrate
Asturias-Arribas et al., 2013	CYP	CV	SPE	Measurement of cocaine	Increase of oxygen current
Liu et al., 2008	CYP2136	CV	AuNP-CH-GCE	Drug sensing	Increase of oxygen current

Table 1. (Continued)

Ref.	CYP species	Technique	Electrode modification	Substrates tested – catalysis	Comments
Fantuzzi et al., 2010	CYP3A4	CV, amperometry, FTIR, HPLC	6-hexanthiol and 7-mercaptohept anoic acid-Au	Quinidine,nifed ipine, alosetron, and ondansetron	Microfluidic system
Wu, 2011b	CYP6A1	CV	DDAB-EPG	Interaction with aldrin and heptachlor	Increase of oxygen current
Baj-Rossi et al., 2012	CYP1A2, 2B6	CV	MWCNT-SPE	Catalysis with O_2 and H_2O_2	Monte Carlo simulations
Shumyantse va et al., 2004	CYP2B4	CV, chronoampero metry	Clay-detergent-GC	Catalysis with aminopyrine, benzphetamine	Increase of oxygen current
Joseph et al., 2003	CYP3A4	QCM, CV, SWV, amperometry electrolysis, product analysis	Au - MPS-PDDA multilayers	Catalysis with verapamil and medazolam	Response time = 15-25s $K_M' = 271 - 1082$ µM.
Xue et al., 2013	CYP3A4	CV	Carbon nanofiber film electrode	Testosterone, quinidine	Increase of oxygen current on substrate addition, effect of inhibitors, DET
Yoshioka et al., 2013	Microsomes CYP3A4	CV	ITO film	Testosterone	Increase of oxygen current on substrate addition, effect of inhibitors
Baj-Rossi et al., 2014	CYP1A2	CV	MWCNT SPE	Continuous monitoring naproxen	LOD=16 µM background subtracted reduction peak

AuNP-CH:Chitosan capped gold nanoparticles, BSA:Bovine serum albumin, DDAB:didodecyldimethylammonium bromide, DET: Direct electron transfer, EPG: Edge-plane pyrolytic graphite electrode, GCE: glassy carbon electrode, LbL: Layer-by-layer, MWCNT: Multi-walled carbon nanotube, PDDA: poly(diallyldimethylammonium chloride), PGE: pyrolytic graphite electrode, SPE:Screen printed electrode.

Different ways to indicate the catalytic substrate conversion by electrochemical or optical transducers have been derived from the reaction cycle (Denisov et al., 2005).

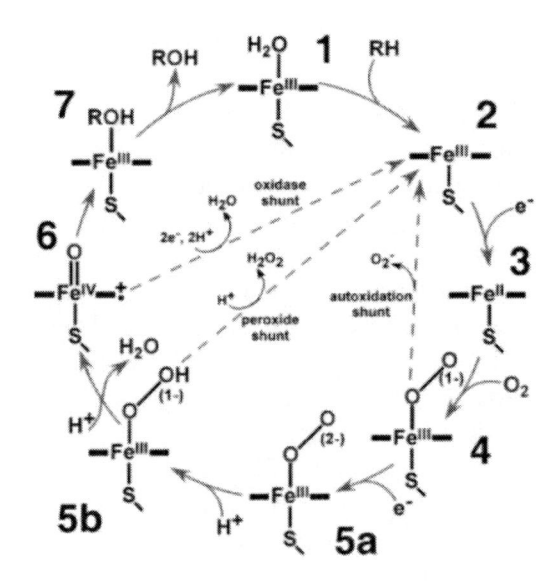

Figure 1. Catalytical cycle of cytochrome P450s (Denisov et al., 2005). Copyright (2013) American Chemical Society.

(i) Indication of a reaction product: The conversion of aromatic substances in CYP catalysed reactions often generates electroactive products. The first CYP based sensor using liver microsomes was already presented in 1978 (Renneberg et al., 1978). It indicated electrochemically the formation of p-aminophenol-the product of NADPH dependent hydroxylation of aniline. The indication of such a product on addition of the cosubstrates NAD(P)H or peroxide is the most direct way for the development of substrate indicating CYP sensors (Renneberg et al., 1978). However, the specificity is limited and an identification of the product by mass spectrometry is therefore suggested.

(ii) Consumption of the cosubstrates NAD(P)H, oxygen or peroxide: The majority of CYP enzymes obtains the two electrons for the activation of oxygen via a NAD(P)H-dependent reductase. Therefore, the conversion of substrates can be monitored through measuring oxygen consumption with the Clark-type oxygen electrode. Also oxygen-mediated fluorescence quenching is a feasible tool for the measurement of the oxygen consumption (Wodnicka et al., 2000). The depletion of NAD(P)H can be followed by tracing the absorbance at 340 nm or by the fluorescence of the oxidised cofactor. However, the parasitic consumption of oxygen and NAD(P)H by both the formation of reactive oxygen species (so called uncoupling reaction of the CYP reaction cycle without product formation) and the presence of other NAD(P)H dependent enzymes are superimposed on the substrate conversion resulting in values too high for the substrate concentration. Uncoupling of the reaction can be followed by measurement of the generation of reactive oxygen species (ROS) using amperometric electrodes for superoxide or peroxide. For good substrate-enzyme pairs only low generation of ROS is expected which reflects tight coupling.

Substrate conversion by the peroxygenase activity can be indicated from the decrease of the peroxide concentration. This principle is well established for peroxidase-based enzyme electrodes (Yu et al., 2006) and is also suitable for the peroxide-dependent CYP catalysis. However, side reaction can contribute to the overall peroxide consumption, thus control measurements in the presence of a CYP inhibitor are useful.

(iii) Mediated cathodic reduction of the protein redox center of CYP: The electrons which are required for substrate conversion may be supplied from a redox electrode via a soluble mediator with a sufficiently low redox potential like viologenes. However, molecular oxygen also reacts with these potent reductants, thus consuming in a "parasitic" reaction, by which the reduced mediator forms hydrogen peroxide. Progress has been achieved by using Co^{2+} sepulchrate as a mediator which reacts only slowly with the ambient oxygen (Estabrook et al., 1996). Covalent binding of the mediator to the protein generates redox relays for fast heterogeneous electron transfer as has been demonstrated for riboflavin-modified CYP (Shumyantseva et al., 2000).

(iv) Electron transfer via redox proteins: The supply of the electrons via the natural electron donating proteins-CYP reductase or the respective iron sulphur protein is the method closest to the physiological conditions. It solves the problem of the "storage" of the TWO electrons for the activation of oxygen. The coupling of liver microsomes via electron transfer to the CYP reductase and de-novo designed redox systems and genetic chimeras (Gilardi et al. 2001; Wong and Schwaneberg, 2003) are straight forward approaches to mimic the pathway of drug metabolism (Yarman et al., 2013).

(v) Direct electron transfer to the heme group: The direct (mediatorless) electron supply from a mercury electrode to the redox-active group of the CYP was established by us (Scheller et al., 1977) in parallel with the development of the 'promoted' direct electron exchange of modified electrodes for c-type cytochromes (Eddowes and Hill, 1977; Yeh and Kuwana, 1977). A whole spectrum of metal electrodes, self-assembled monolayers, nano-materials such as different types of carbon-nanotubes, screen-printed and micro-fabricated electrodes are increasingly applied in CYP sensors. In recent years profit has been taken from the use of protein engineering which includes the introduction of specific attachment regions for surface binding and redox relays.

CYP SENSORS FOR PERSONALISED MEDICINE

Despite intensive drug research and expensive clinical trials at present almost 90 percent of the drugs show a therapeutic effect for only 40 percent of treated patients (Baj-Rossi et al., 2011). A therapy which is based on individual drug metabolising activity would avoid unnecessary medication and reduce costs. This "companion diagnostics" has been established for the administration of Warfarin-containing pharmaca to lower the risk of thrombosis or strokes. The rate of metabolism of Warfarin is primarily regulated by variations in CYP activity. In order to characterise the individual pharmacokinetics of a drug it is essential to measure the metabolic activity of the enzyme (Panicco et al., 2011). The characterization of the kinetics of drug conversion by a given CYP isoenzyme and its polymorphic variants has been addressed by forming fusion proteins bearing P450 and flavodoxin, which were then bound to gold electrodes modified with a self-assembled monolayer. The ability to engineer, express and purify a whole spectrum of CYP isoenzymes and polymorphic variants has enabled the successful development of CYP-arrays for in-vitro drug testing (Panicco et al., 2011).

CHALLENGES/COMMENTS

In spite of more than 30 years of research, interpretation of the reaction mechanism of signal generation in many published CYP-based sensors is complicated by the fact that substrate conversion requires oxygen whilst the reduction of the CYP heme is mashed by the cathodic oxygen reduction. The cathodic reduction of the ferric CYP or the mediators requires a potential at which normally also oxygen is reduced. Most probably, both reduction of the prosthetic group and of oxygen proceed in parallel at the electrode. Therefore the electrochemical injection of the electrons is accompanied under "reaction conditions" by the generation of hydrogen peroxide and the reaction may follow the "peroxide shunt". Therefore the effect of catalase should be checked.

An increase of the cathodic current for oxygen reduction on addition of the substrate is not a serious criterion for electrode -supported substrate conversion. Because for each oxygen molecule only one molecule of a potentially electroactive product can be formed, a decrease of the oxygen current on addition of the substrate is plausible. Therefore the bulk oxygen concentration should be carefully controlled in order to avoid fluctuations on the addition of the substrate which is dissolved in an organic solvent.

The influence of CYP inhibitors can be a useful diagnostic criterium. However, complete suppression of the catalytic oxygen reduction in the presence of a substrate by heme-binding inhibitors is not a proof for electrode-driven substrate conversion in the absence of the inhibitor. Only the identification of the reaction product is the "gold standard".

REFERENCES

Asturias-Arribas, L., Alonso-Lomillo, M.A., Domínguez-Renedo, O., Arcos-Martínez, M.J., (2013) Electrochemical determination of cocaine using screen-printed cytochrome P450 2B4 based biosensors, *Talanta* 105, 131-134.

Baj-Rossi, C., De Micheli, G., Carrara, S., (2011) P450-Based Nano-Bio-Sensors for Personalized Medicine, in: P.A. Serra (Ed.), Biosensors-Emerging materials and applications, INTECH, DOI: 10.5772/16328.

Baj-Rossi, C., De Micheli, G., Carrara, S., (2012) Electrochemical detection of anti-breast-cancer agents in human serum by cytochrome P450-coated carbon nanotubes, *Sensors* 12, 6520-6537.

Baj-Rossi, C., Rezzonico Jost, T., Cavallini, A., Grassi, F., De Micheli, G., Carrara, S., (2014) Continuous monitoring of Naproxen by a cytochrome P450-based electrochemical sensor, *Biosens. Bioelectron.* 53, 283-287.

Carrara, S., Cavallini, A., Erokhin, V., De Micheli, G., (2011) Multi-panel drugs detection in human serum for personalized therapy, *Biosens. Bioelectron.* 26 (2011) 3914.

Denisov, I.G., Makris, T.M., Sligar, S.G., Schlichting, I., (2005) Structure and chemistry of cytochrome P450, *Chem. Rev.* 105, 2253-2277.

Eddowes, M.J., Hill, H.A.O., (1977) Novel method for the investigation of the electrochemistry of metalloproteins: cytochrome *c*, *J. Chem. Soc., Chem. Commun.* 21, 771b-772.

Estabrook, R.W., Faulkner, K.M., Shet, M.J., Fisher, C.W., (1996) Application of electrochemistry for P450-catalyzed reactions, in: E. F. Johnson and M.R. Waterman (Eds.), Methods in Enzymology, Vol:272, Elsevier, 1996, p.44-51.

Fantuzzi, A., Capria, E., Mak, L.H., Dodhia, V.R., Sadeghi, S.J., Collins, S., Somers, G., Hug, E., Gilardi, G., (2010) An electrochemical microfluidic platform for human P450 drug metabolism profiling, *Anal. Chem.* 82 (2010) 10222-10227.

Gilardi, G., Fantuzzi, A., Sadeghi, S.J., (2001) Engineering and design in the bioelectrochemistry of metalloproteins, *Curr. Opin. Struct. Biol.* 11, 491-499.

Girhard, M., Kunigk, E., Tihovsky, S., Shumyantseva, V.V., Urlacher, V.B., (2013) Light-driven biocatalysis with cytochrome P450 peroxygenases, *Biotechnol Appl Biochem.* 60, 111-118.

Hung, K.W., Suen, M.F., Chen, Y.F., Cai, H.B., Mo Z.X., Yung, K.K., (2012) Detection of water toxicity using cytochrome P450 transgenic zebrafish as live biosensor: for polychlorinated biphenyls toxicity, *Biosens. Bioelectron.* 31, 548-553.

Iwuoha E.I., Joseph S., Zhang Z., Smyth M.R., Fuhr, U., Ortiz de Montellano P.R., (1998) Drug metabolism biosensors: Electrochemical reactivities of cytochrome CYP101 immobilised in synthetic vesicular systems, *J. Pharm. Biomed. Anal.* 17, 1101-1110.

Iwuoha, E.I., Wilson, A., Howel, M., Mathebe, N.G.R., Montane-Jaime, K., Narinesingh, D., Gueseppi-Elie, A., (2004) Cytochrome P-4502D6 (CYP2D6) bioelectrode for fluoxetine, *Anal. Lett.* 37, 929-941.

Joseph, S., Rusling, J.F., Lvov, Y.M., Fredberg, T., Fuhr, U., (2003) An amperometric biosensor with human CYP3A4 as a novel drug screening tool, *Biochem. Pharmacol.* 65, 1817-1826.

Krishnan, S., Wasalathanthri, D., Zhao, L.L., Schenkman, J.B., Rusling, J.F., (2011) Efficient bioelectronic actuation of the natural catalytic pathway of human metabolic cytochrome P450s, *J. Am. Chem. Soc.* 133, 1459-1465.

Liu, S.Q., Peng, L., Yang, X.D., Wu, Y.F., He, L. (2008) Electrochemistry of cytochrome P450 enzyme on nanoparticle-containing membrane-coated electrode and its applications for drug sensing, *Anal. Biochem.* 375, 209-216.

Panicco, P., Dodhia, V.R., Fantuzzi, A., Gilardi, G., (2011) Enzyme-based amperometric platform to determine the polymorphic response in drug metabolism by cytochromes P450, *Anal. Chem.* 83, 2179-2186.

Rabe, K.S., Gandubert, V.J., Spengler, M., Erkelenz, M., Niemeyer C.M. (2008) Engineering and assaying of cytochrome P450 biocatalysts, *Anal. Bioanal. Chem.* 392, 1059-1073.

Renneberg, R., Scheller, F., Ruckpaul, K., Pirrwitz, J., Mohr, P., (1978) NADPH and H_2O_2-dependent reactions of cytochrome P-450LM compared with peroxidase catalysis, *FEBS Lett.* 96, 349-353.

Scheller, F., Renneberg, R., Strnad, G., Pommerening, K., Mohr, P., (1977) Electrochemical aspects of cytochrome P-450 system from liver microsomes, *Bioelectrochem. Bioenerg.* 4 (1977) 500.

Shumyantseva, V.V., Ivanov, Y.D., Bistolas, N., Scheller, F.W., Archakov, A.I., Wollenberger, U., (2004) Direct electron transfer of cytochrome P450 2B4 at electrodes modified with non-ionic detergent and colloidal clay nanoparticles, *Anal. Chem.* 76, 6046-6052.

Shumyantseva, V.V., Bulko, T.V., Bachmann, T.T., Bilitewski, U., Schmid, R.D., Archakov, A.I., (2000) Electrochemical reduction of flavocytochromes 2B4 and 1A2 and their catalytic activity, *Arch. Biochem. Biophys.* 377, 43-38.

Shumyantseva, V.V., Bulko, T.V., Usanov, S.A., Schmid, R.D., Nicolini, C., Archakov, A.I., (2001) Construction and characterization of bioelectrocatalytic sensors based on cytochromes P450, *J. Inorg. Biochem.* 87, 185-190.

Wodnicka, M., Guarino, R.D., Hemperly, J.J., Timmins, M.R., Stitt, D., Pitner, J.B., (2000) Novel fluorescent technology platform for high throughput cytotoxicity and proliferation assays, *J. Biomol. Screen.* 5, 141-152.

Wong, T.S., Schwaneberg, U., (2003) Protein engineering in bioelectrocatalysis, *Curr. Opin. Biotech.* 14, 590-596.

Wu, Y., Liu, X., Zhang, L., Wang, C., (2011a) An amperometric biosensor based on rat cytochrome P450 1A1 for benzo[a]pyrene determination, *Biosens. Bioelectron.* 26, 2177-2182.

Wu, Y., (2011b) Direct electrochemistry of cytochrome P450 6A1 in mimic bio-membrane and its application for pesticides sensing, *Sensor. Actuat. B-Chem.* 156, 773-778.

Xue, Q., Kato, D., Kamata. T., Guo, Q., You, T., Niwa, O., (2013) Human cytochrome P450 3A4 and a carbon nanofiber modified film electrode as a platform for the simple evaluation of drug metabolism and inhibition reactions, *Analyst* 138, 6463-6468.

Yu, D. Blankert, B., Bodoki, E., Viré, J.-C., Sandulescu, R., Nomura, A. Kauffmann, J.-M. (2006) Amperometric biosensors based on horseradish peroxidase-immobilised magnetic microparticles, *Sensor. Actuat. B-Chem.* 113, 749-754.

Yarman, A., Wollenberger, U., Scheller, F.W., (2013) Sensors based on cytochrome P450 and CYP mimicking systems, *Electrochim. Acta* 110, 63-72.

Yeh, P., Kuwana, (1977) Reversible electron reaction of cytochrome *c*, *Chem. Lett.* 10, 1145-1148.

Yoshioka, K., Kato, D., Kamata, T., Niwa, O., (2013) Cytochrome P450 modified polycrystalline indium tin oxide film as a drug metabolizing electrochemical biosensor with a simple configuration, *Anal. Chem.* 85, 9996-9999.

In: Cytochrome P450 Enzymes
Editor: Jian Wu

ISBN: 978-1-61942-209-4
© 2014 Nova Science Publishers, Inc.

Chapter 2

CYTOCHROME P450-MEDIATED TOXICITY OF THERAPEUTIC DRUGS

Mariana Matias [1,2], Catarina Canário [1],
Samuel Silvestre[1,2], Amílcar Falcão[2,3]
and Gilberto Alves[1,2]*

[1]CICS-UBI – Health Sciences Research Centre,
University of Beira Interior, Covilhã, Portugal
[2]CNC – Centre for Neuroscience and Cell Biology,
University of Coimbra, Coimbra, Portugal
[3]Department of Pharmacology, Faculty of Pharmacy,
University of Coimbra, Pólo das Ciências da Saúde, Coimbra, Portugal

ABSTRACT

Cytochrome P450 (CYP450) system is a superfamily of heme-containing monooxygenase enzymes and represents one of the most extensively studied enzymatic systems worldwide. The high interest on CYP450 enzymes reflect not only its importance in the metabolic detoxication of drugs and other xenobiotics from the body, but also its

* Corresponding Author address: Faculty of Health Sciences, University of Beira Interior; CICS-UBI – Health Sciences Research Centre, University of Beira Interior; Av. Infante D. Henrique, 6200-506 Covilhã, Portugal; Phone: +351 275329002 / Fax: +351 275329099; Email: gilberto@fcsaude.ubi.pt

significant role in the aetiology of some diseases and in metabolic-mediated toxification processes (bioactivation).

Although CYP450 enzymes are most highly expressed in the liver, they are also expressed in many other tissues such as small intestine and kidneys, which also contribute decisively to determine the fate of the drugs in the body. Indeed, the metabolism of drugs involves numerous biotransformation reactions which are essentially chemical processes mainly mediated by enzymes and usually aiming drug elimination through inactive and more easily excretable drug metabolites. Unfortunately, for almost all drug classes, those biotransformation reactions may also be responsible for toxic unwanted effects that, in some circumstances, can lead to drastic and irreversible consequences to human health and even death. In fact, over the years, several research works have described toxicological effects of several clinically used drugs due to CYP450-mediated metabolic conversion to toxic metabolites and, in some cases, the molecular mechanisms associated have also been investigated. For this reason, during the development of new drugs it is of major relevance to identify potential chemical groups in their structure that can potentially be associated with toxic effects (toxicophoric groups) and, when possible, they should be changed to bioequivalent groups.

As drug metabolism and/or toxicity are mainly mediated by CYP450 enzymes, this chapter highlights recent and relevant drug examples in this context including the molecular mechanisms of toxicity and the functional groups responsible for the toxic effects, aiming to contribute for the development of safer drugs.

Keywords: Cytochrome P450, Drug metabolism, Drug bioactivation, Toxicity

INTRODUCTION

Metabolism is one of the major determinants of the fate of drugs in the body, as it determines their pharmacokinetic properties, and thus, their efficacy and toxicity. Drug metabolism (biotransformation) involves a large variety of chemical reactions, mainly mediated by enzymes, which usually lead to the conversion of drugs into inactive and more readily excretable compounds (metabolites). Therefore, drug metabolism is mainly considered a detoxification pathway. However, under certain circumstances, even at the usual therapeutic doses, harmful reactive metabolites more toxic than the parent drugs can also be generated. [1, 2]

The majority of drug biotransformation reactions occur inside of the cells often involving multiple simultaneous metabolic pathways.[3] In this context, in a comprehensive review, Croom [4] recently pointed out that most xenobiotics are cleared through multiple enzymes and metabolic pathways, and the relationship between several factors (e.g., chemical concentration, enzyme affinity and quantity, and cofactor availability) determines which is the major metabolic pathway in a given subject.

In spite of the aforementioned, drug metabolism is traditionally described as two or three sequential phases: Phase I (also known as *functionalization phase*), Phase II (also called *conjugation phase*) and Phase III (export from the cell by transporter proteins).

Hence, it is important to have in mind, in all circumstances, that metabolism is a dynamic process and that the cells have important functions (e.g., enzymes and cofactors) for biotransformation of both endogenous and exogenous (xenobiotics) compounds. [2,3,5] However, it is also true that lipophilic xenobiotics (including drugs) are often firstly metabolized by Phase I enzymes aiming to make them more polar and thus allowing subsequent conjugation reactions.

The Phase I metabolic functionalization reactions include oxidation (by far, the most important), reduction and hydrolysis and aims to introduce (directly or by a deprotection reaction) a functional group (e.g. hydroxyl or amino group) that can react with an activated endogenous molecule in Phase II. The Phase II enzymes essentially act as transferases (e.g., uridine diphosphate glucurono-syltransferases, *N*-acetyltransferases, glutathione *S*-transferases), which can directly interact with xenobiotics, catalyzing their conjugation with an endogenous molecule (e.g., glucuronic acid, acetate, glutathione); however, more frequently than the direct reaction with parent drugs, they interact with the metabolites produced by Phase I enzymes. Besides some degree of passive diffusion of the newly formed Phase II compounds to the outside of cells, a number of Phase III transporters can also actively pump the more polar conjugated xenobiotics through the cell membrane. [3-5]

The cytochrome P450 (CYP450) system comprises an enzyme superfamily of heme-containing monooxygenases, which are by far the most important Phase I metabolizing enzymes and participate in a large number of xenobiotic metabolic reactions. Indeed, CYP450 enzymes are found in all domains of life such as in bacteria, fungi, plants and mammals. [6-8]

Cytochrome P450

CYP450 was firstly named in 1961 to identify the cellular chromophore/pigment (P) that has a spectrophotometric absorption peak at 450 nm when reduced and bound to carbon monoxide. [9] Historically, it is also important to highlight the research works published nearly 50 years ago by Omura and Sato, [10, 11] which decisively contributed to the identification and characterization of CYP450 enzymes, representing a milestone in the area of drug metabolism. Afterwards, the purification and crystallization of different CYP450 enzymes have been successfully achieved. In 1974, Yu and collaborators [12] described the purification and crystallization of $CYP450_{cam}$ obtained from camphor-induced cells of *Pseudomonas putida*. More recently, in 2000, the crystallization of the first mammalian CYP450 structure (rabbit CYP2C5) was published by Cosme and Johnson. [13, 14] Up to this time, the structural models for human CYP450 enzymes were derived from the data obtained using the distantly related bacterial CYP450s. [15] Hence, this achievement was also an important landmark on drug (xenobiotic) metabolism, opening new perspectives for modeling other mammalian CYP450 enzymes and providing novel insights to aid in the discovery of more effective and safer drugs. In spite of the fact that only a few years passed between the determination of the crystal structure of the rabbit CYP2C5 isoenzyme, [13,14] nowadays the crystal structures of several human CYP450 isoenzymes such as CYP2C9, [16] CYP2C8, [17] CYP3A4, [18] and CYP2D6 are also available. [19]

CYP450 enzymes are ubiquitous in nature and are found in almost all mammalian tissues such as brain, kidney, heart, intestine, lung, nasal and tracheal mucosa, adrenal gland, gonads, skin, and many other tissues. However the highest abundance and largest number of individual CYP450s is found in the liver. [3, 7] CYP450 enzymes represent nearly 2% of the total human liver microsomal protein (0.3-0.6 nmol of total CYP450 per mg of microsomal protein), but this content is much lower in other tissues. [7] CYP450 enzymes are traditionally thought to be localized, at the sub-cellular level, in the smooth endoplasmatic reticulum (microsomal CYP450s). Indeed, the current drug metabolism and drug toxicity studies mainly rely on the microsomal CYP450 pool, which undoubtedly is the major system responsible for the metabolic activity of the tissues. However, new findings have also shown that a number of inducible and constitutive CYP450s is also present in the mitochondrial compartments. [3, 20]

Today it is well-recognized that CYP450 represents one of the most extensively studied enzymatic systems worldwide and there are no doubts that it is the major xenobiotic (drug) metabolizing system. [21, 22] In the human genome there are 57 genes and more than 59 pseudogenes divided among 18 families of CYP450 genes and 43 subfamilies. [23] Obviously, due to the numerous CYP450 enzymes, a nomenclature based on their sequence similarity has been established to classify them in the corresponding families and subfamilies. Briefly, CYP450 isoenzymes are designated by the capital root letters CYP for cytochrome P450, followed by an Arabic number indicating the corresponding family (at least 40% amino acid sequence homology), a capital letter indicating the subfamily (at least 55% amino acid sequence homology), and another Arabic number for the individual gene. [24] Although 57 putatively functional human CYP450s have been identified, only about a dozen enzymes, belonging to the CYP1, CYP2, and CYP3 families, are responsible for the biotransformation of most foreign substances including 70-80% of all therapeutic drugs. [22] Many other CYP450 isoenzymes are involved in the metabolism of sterols, fatty acids, eicosanoids and vitamins, which are important in normal physiology, whereas the role of other CYP450 isoenzymes is still unknown. [25] Madanayake and colleagues [26] suggested that approximately one-quarter out of the 57 distinct human CYP450s are classified as orphans with little or no knowledge about their substrates and physiological significance.

Specifically, the CYP450 isoenzymes primarily responsible for the metabolism of therapeutic drugs in the human liver mostly include CYP1A2, CYP1B1, CYP2A6, CYP2B6, CYP2C8, CYP2C9, CYP2C19, CYP2D6, CYP2E1, CYP3A4 and CYP3A5. [7, 27] Of all CYP450 enzymes, the CYP3A4 is the most abundant isoform and is the main metabolizing system of therapeutic drugs and also a very large amount of other xenobiotics. [7] The human enzymes of the CYP3A subfamily, mainly CYP3A4 and CYP3A5, mediate the metabolism of approximately half all marketed drugs and thus play a critical role in drug metabolism. [28] Moreover, a well-known particularity of CYP450 enzymes is their ability to metabolize a large spectrum of structurally diverse xenobiotics. In fact, there are comprehensive reviews in literature that contain exhaustive lists of xenobiotics that are CYP450 substrates. [7, 28, 29].

Metabolism represents the main route of elimination for near 70% of the currently available drugs in the clinical practice. Bearing in mind the lack of substrate selectivity and the pivotal role of CYP450 enzymes in drug metabolism, catalyzing numerous oxidative Phase I reactions, it is not surprising that these enzymes play a key role in relevant drug-drug interactions. In addition, the activity of CYP450 enzymes is influenced by numerous genetic and environmental factors, making drug metabolism highly variable, which inevitably impacts on drug development, clinical drug therapy and on the sensitivity to xenobiotics. [7, 30, 31]

In contrast to CYP450 enzymes involved in the metabolism of sterols, the expression levels and functional metabolic activity of drug-metabolizing CYP450 enzymes varies widely between individuals. Particularly, the genes encoding CYP450 enzymes are highly polymorphic (e.g., CYP2C9, CYP2C19, CYP2D6 and CYP2B6) and this may lead to therapeutic inefficacy in ultrarapid metabolizers, or toxicity in poor metabolizers for many drugs administered at a standard dose. [25, 30, 31]

Despite the key function of CYP450 enzymes in the inactivation and elimination of a large variety of drugs from the body, many CYP450-mediated drug oxidative reactions are partially uncoupled, leading to the generation of toxic reactive metabolites. These reactive species may covalently bind to hepatic or extra-hepatic macromolecules causing toxicity. [1, 32] In fact, paradoxically, many of the adverse effects of drugs may be attributable to the CYP450 enzymes, being the CYP450-mediated drug bioactivation to reactive metabolites often reported as the first step in many adverse drug reactions. [1, 33] Hence, the primary focus of this chapter is the discussion of relevant examples of therapeutic drugs for which their toxic effects are, at least partially, ascribed to CYP450-mediated toxicity. This collection of information may be useful to better understand the CYP450-dependent bioactivation and to aid in the design and development of safer drugs.

CYP450-MEDIATED BIOACTIVATION OF DRUGS

Due to its relevance, over the years, drug metabolism has become a main area of interest, having particular importance in medicinal chemistry, pharmacology, therapeutics and toxicology. This situation results from the fact that, as previously referred, metabolism can influence the deactivation, activation, detoxification and toxification of most drugs. For this reason, it is of major relevance to know the metabolic way, as well as the potential toxicity as early as possible in the development of new drugs. [34-36].

Toxic effects can be directly originated by the chemical compound or can be the result of its metabolism (indirect toxicity), either from a reactive intermediate or a primary or even a final metabolite. Thus, in addition to the knowledge of the metabolic way, it is also important to disclose the mechanisms involved in the toxic effects caused by chemicals, which commonly are directly related to their metabolism. [34, 35, 37]

In the context of toxicity mediated by metabolic reactions, bioactivation is defined as biotransformation(s) that produce toxic metabolites, usually electrophilic structures, which are able to react with tissue nucleophiles, namely protein lysine amines and cysteine thiols as well as DNA guaninic nitrogen-7. These changes in endogenous structures can lead to a functional cellular dysfunction or even a toxic response. [36, 38, 39]

Bioactivation reactions are in most cases Phase I metabolic reactions. However, in some cases, Phase II transformations can also be involved in toxicity. An example of the latter situation is the sulfation reaction, which can introduce a good leaving group, allowing the formation of electrophilic carbocations. [39, 40]

The most important metabolic systems involved in bioactivation reactions are CYP450 and peroxidase enzymes, producing either electrophilic or radical metabolites. Of these, the CYP450 system is frequently considered the most relevant, generating primarily electrophiles and less frequently free radicals. [36] This system is usually involved in oxidative reactions of xenobiotics and endogenous compounds but in anaerobic conditions, reductive transformations can also be catalyzed by some CYP450 enzymes. CYP450 monooxigenase is the most important oxidizing metabolic system, is mainly located in the liver and is implicated in *C*-, *N*- and *S*-oxidations, *N*-, *O*- and *S*-dealkylations, deaminations and other reactions. After binding the substrate, its catalytic mechanism involves the activation of oxygen helped by the addition of two electrons, leading to the formation of an electrophilic and reactive oxene. This compound reacts with the substrate, which is oxidized, usually through transient radical intermediates (Figure 1). [6, 41] The oxidized product can be more toxic than its precursor compound and/or suffer further metabolic transformation. Free radicals that are formed in CYP450 oxidative reactions, when non-efficiently controlled, may react at the site of their formation, leading to inhibition and/or inactivation of the enzyme, or can be released and lead to lipid peroxidation and cellular dysfunction/destruction. [42]

Common oxidative transformations involved in toxicity reactions [41] usually include saturated and unsaturated carbon-oxidations (examples in Figure 2) as well as heteroatom oxidations, mainly nitrogen atoms. In this context, saturated carbon oxidation originating hydroxylated derivatives, which is a very common metabolic reaction, can sometimes be especially problematic if the transformation occurs in the α-position to a heteroatom.

[39, 40] This reaction can allow further reactions such as eliminations in halogenated compounds (Figure 2) or cleavages in some amines that lead to reactive metabolites. An example is a tertiary amine that, after an allylic hydroxylation, affords a carbinolamine which can suffer a C-N bond cleavage originating a secondary amine and a reactive aldehyde (Figure 2).

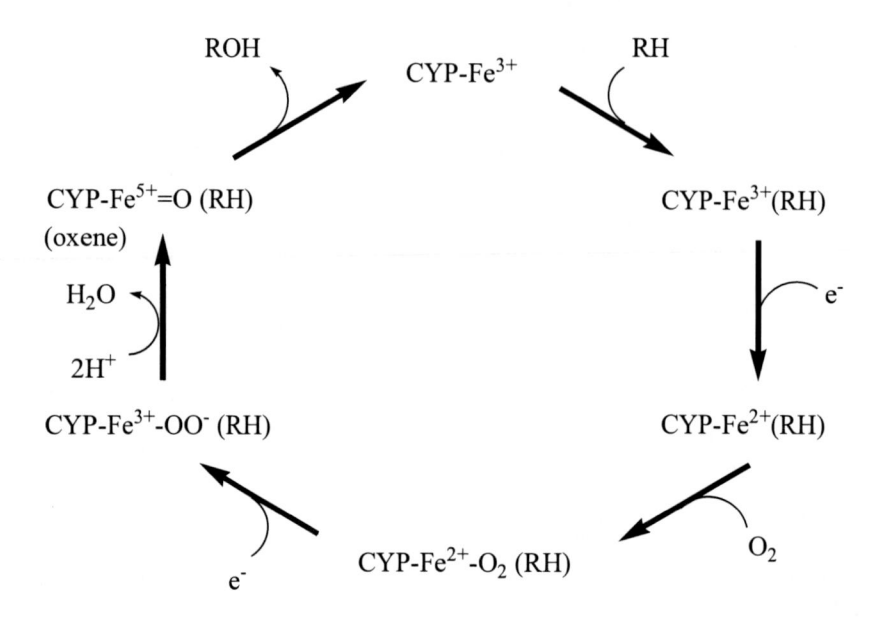

Figure 1. Basic CYP450 monooxygenase catalytic cycle.

Epoxides and arene oxides are usually considered responsible for the toxicity of unsaturated compounds and aromatic chemicals. [39-41] These structures are also usually formed through oxidation mediated by CYP450 and can then suffer an attack by cellular nucleophiles (Nu) leading to the epoxide ring opening (Figure 2) and a structural change in the involved cellular components. Primary and secondary amines are functional groups commonly found in drug structures. These amines can, under certain conditions, be N-oxidized by CYP450 oxidases into the corresponding hydroxylamines, which can directly be mutagenic and carcinogenic or can react and originate other toxic compounds [39, 40].

Oxidative stress may also be involved in the toxicity effects of some drugs and/or their metabolites, such as quinones or analogues, as well as nitroaromatic compounds. These compounds can suffer one-electron reduction, allowing the production of radical structures, which can be reoxidized to their parent structure accompanied by the reduction of oxygen to superoxide anion. This compound, as well as other reactive oxygen species (ROS) that can subsequently be formed, such as hydrogen peroxide and hydroxyl radical, can attack several cellular components leading to their dysfunction/destruction. This effect can be especially exacerbated if the normal ROS degradation systems (e.g. superoxide dismutase, catalase and glutathione peroxidase) are overburdened. [42, 43]

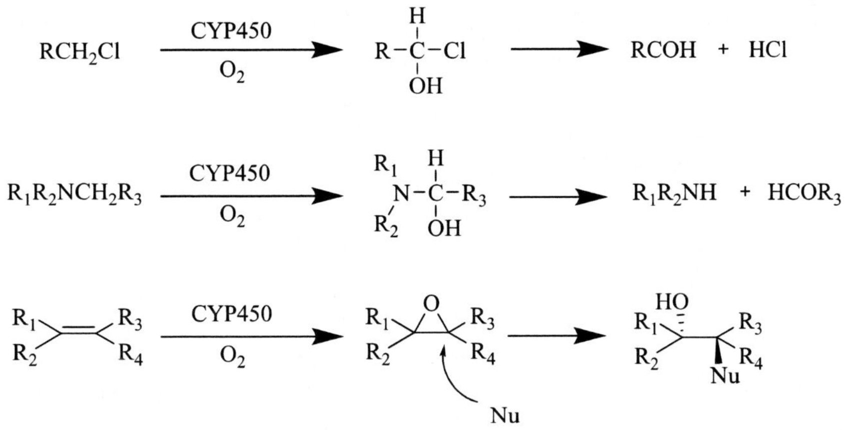

Figure 2. Examples of common oxidative transformations that can be involved in toxicity responses (Nu = nucleophile).

Several compounds, including polyhalogenated, nitro and azo derivatives can also suffer CYP450-mediated reductive transformations which can also lead to the formation of very destructive carbon-, nitrogen- and oxygen-centered radicals, as well as other reactive structures. It is also worth mentioning that although reductive metabolic reactions are less frequent than oxidations, they can have dramatic toxic consequences. [39, 40, 44, 45]

Having in mind these last considerations, it is clear that several functional groups are clearly associated with toxic effects (toxicophoric groups), either directly or as the result of their metabolism. [37, 46] Due to these reasons, it is not surprising that lists of functional groups involved in toxicity, as well as the main bioactivation pathways, are being considered in drug design and development programs. Obviously, functional groups that can, either directly or indirectly, lead to a toxic response should be avoided in drug design. In this context, and aiming at a more rational and successful design/development of drug candidates with a lower susceptibility to form reactive intermediates and a lower toxicity, recently, researchers also created a list of "structural alerts".

This list includes molecular fragments associated with the formation of reactive intermediates and adverse drug reactions, as well as chemical groups that are present in safe drugs. [35] As it is not always possible to avoid the presence of structural groups that are associated with bioactivation, the prevalent strategy is to make a structural change that reduces or completely blocks metabolism. This includes the replacement of the suspect functional group, the use of metabolic blockers (e.g. a halogen), the sterical hindrance of the metabolism and the change to "soft" metabolic groups. [36]

RELEVANT EXAMPLES OF CYP450-MEDIATED DRUG BIOACTIVATION

Acetaminophen

Acetaminophen (Figure 3) is an effective analgesic and antipyretic drug widely used for the relief of mild to moderate pain. While this drug is generally regarded as safe when used at recommended doses, large doses of acetaminophen can cause liver injury and even hepatic failure. The mechanism of acetaminophen-induced hepatotoxicity in humans involves its covalent binding to critical proteins, oxidative stress, mitochondrial dysfunction and nuclear deoxyribonucleic acid (DNA) fragmentation. [47, 48]

At therapeutic doses, acetaminophen is primarily metabolized by glucuronidation and sulfation of its phenolic ring in the liver, with acetaminophen-glucuronide and acetaminophen-sulfate being the main biotransformation products circulating in plasma (Figure 3). [49] However, a small proportion of the drug undergoes oxidation via CYP450 system. [50] Thus, the liver injury caused by acetaminophen appears to be initiated by its CYP450-mediated bioactivation to N-acetyl-$para$-benzoquinoneimine (NAPQI, Figure 3), which is an electrophilic and oxidative reactive metabolite. Although the CYP450 isoenzymes involved in drug metabolism include CYP1A2, CYP2A6 and CYP3A, CYP2E1 is widely recognized as the major isoform implicated in the biotransformation of acetaminophen to NAPQI. [51] This metabolite is efficiently detoxified by glutathione (GSH) conjugation or reduction to its parent compound, acetaminophen.

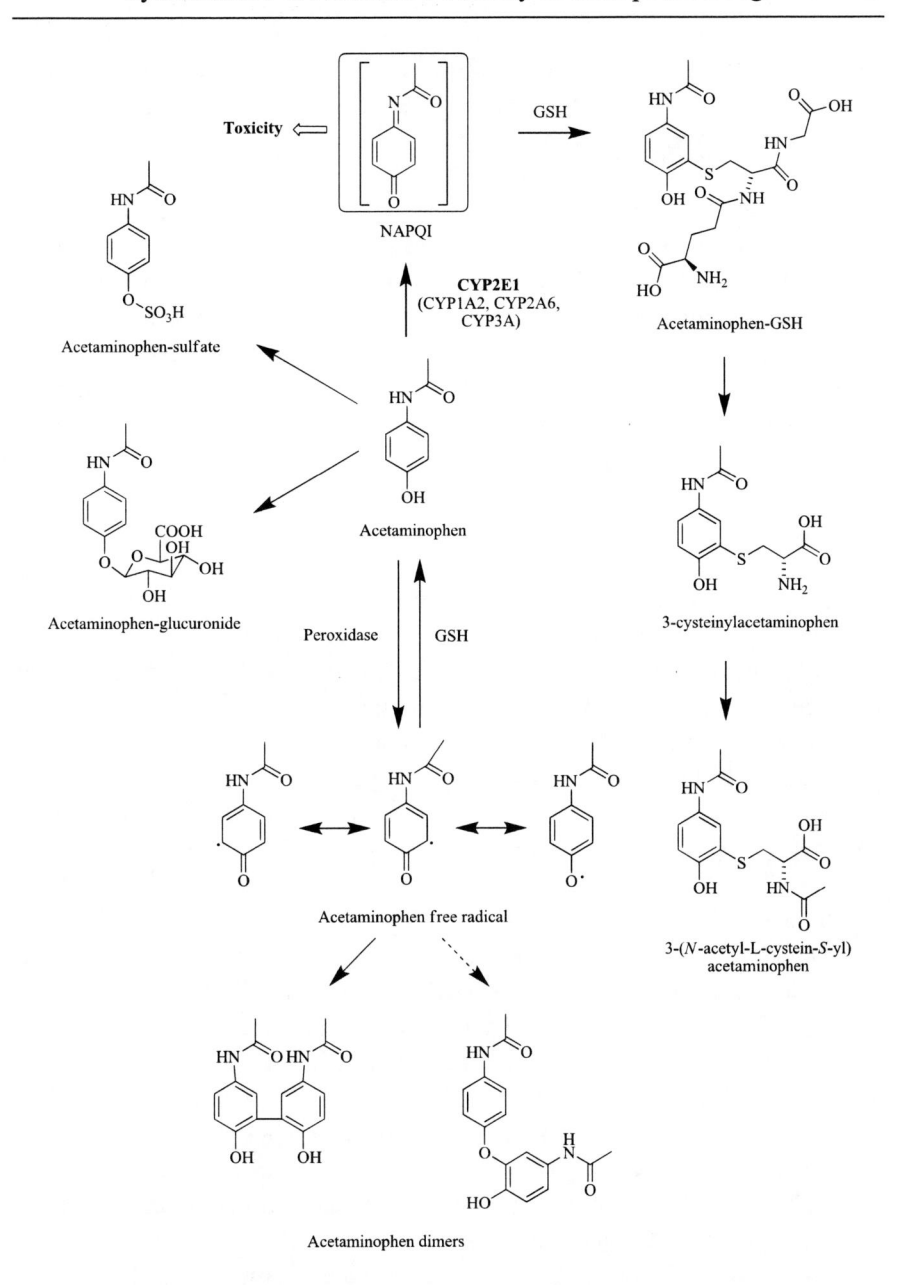

Figure 3. Metabolic pathways of acetaminophen, with production of *N*-acetyl-*para*-benzoquinoneimine (NAPQI), the main reactive intermediate metabolite. [49, 53, 54]

However, after an exposure to very high drug concentrations, hepatic GSH is depleted and can no longer prevent the accumulation of NAPQI, leading to its binding to proteic cysteine sulfhydryls, forming acetaminophen protein adducts (Figure 3). [52] Nevertheless, recent evidence suggests that these adducts can also be formed following lower-dose exposure to acetaminophen. [50]

In addition, cyclooxygenase-1 and cyclooxygenase-2, through their peroxidase function, can generate the acetaminophen phenoxyl free radical by acetaminophen oxidation. In *in vitro* conditions, this radical undergoes either rapid dimerization with another acetaminophen radical, or reduction by GSH to the parent compound (Figure 3) with the consequent formation of GSH disulphide. However, to what extent these dimers are formed *in vivo* and their clinical relevance remains unclear. [53]

Furosemide

Furosemide (Figure 4) is an anthranilic acid derivative and a potent loop diuretic, which is frequently used in the management of fluid retention associated with cardiac, renal, and hepatic failure, as well as in the treatment of hypertension. This drug acts by inhibition of the Na^+-K^+ and $2Cl^-$ co-transporter in the ascending limb of the loop of Henle, which results in the retention of water in the tubule. [55] At therapeutic doses, furosemide is considered non toxic to humans, although it has been associated with jaundice and hypersensitivity. However, there is a potential risk of hepatotoxicity associated with this drug when administered in large doses to patients suffering acute and chronic renal failure. [56]

Research with microsomal preparations confirmed that CYP450 isoenzymes are involved in the overall metabolism of furosemide in rats, namely CYP2C11, CYP2E1, CYP3A1, and CYP3A2, [57] and that the bioactivation of the furan ring represents the main mechanism of metabolic activation and hepatotoxicity of this drug. [58] The major metabolic routes include glucuronidation, *N*-dealkylation to 2-amino-4-chloro-5-sulfamoyl-benzoic acid, and furan ring oxidation to originate a GSH conjugate and a γ-ketocarboxylic acid metabolite. This last compound can require the formation of a γ-ketoenal intermediate (Figure 4). [56,59]

The epoxide and γ-ketoenal intermediates formed in furosemide bioactivation (Figure 4) play an important role in the production of toxic metabolites. The electrophilicity of the furosemide-epoxide metabolite results in attacks by nucleophilic macromolecules, leading to the formation of covalently binding products and disruption of calcium homeostasis, that can produce several side effects. On the other hand, the γ-ketoenal furosemide metabolite can be trapped, forming an *N*-acetylcysteine/*N*-acetyl lysine furosemide adduct (Figure 4). [58]

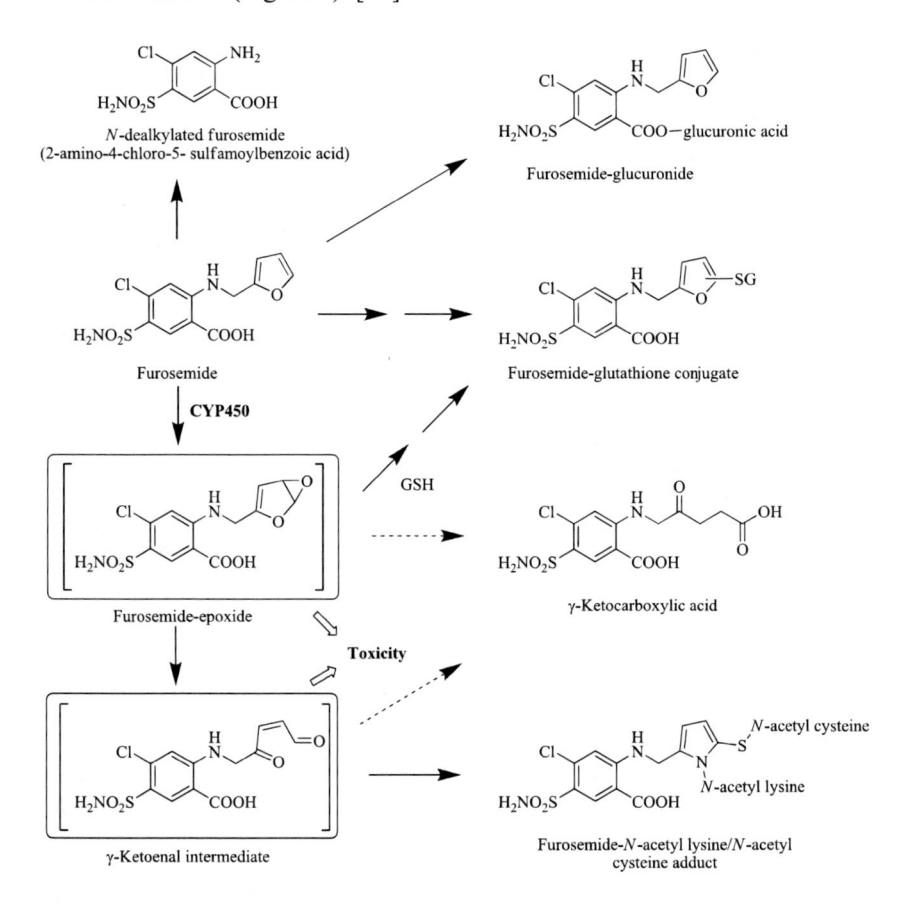

Figure 4. Major metabolic pathways of furosemide, showing the γ-ketocarboxylic acid and γ-ketoenal metabolites formed via an epoxide intermediate. [56, 58]

Diclofenac

Diclofenac (Figure 5) is a nonsteroidal anti-inflammatory drug that is widely used in the treatment of mild to moderate pain and of a variety of rheumatic disorders. It is a phenylacetic acid derivative, with the carboxylic acid moiety being involved in the production of reactive glucuronide metabolites. In addition, diclofenac has a diphenylamine backbone, with the secondary amine contributing to its uncoupling activity in mitochondria, besides being involved in the production of oxidoreductive stress. [60]

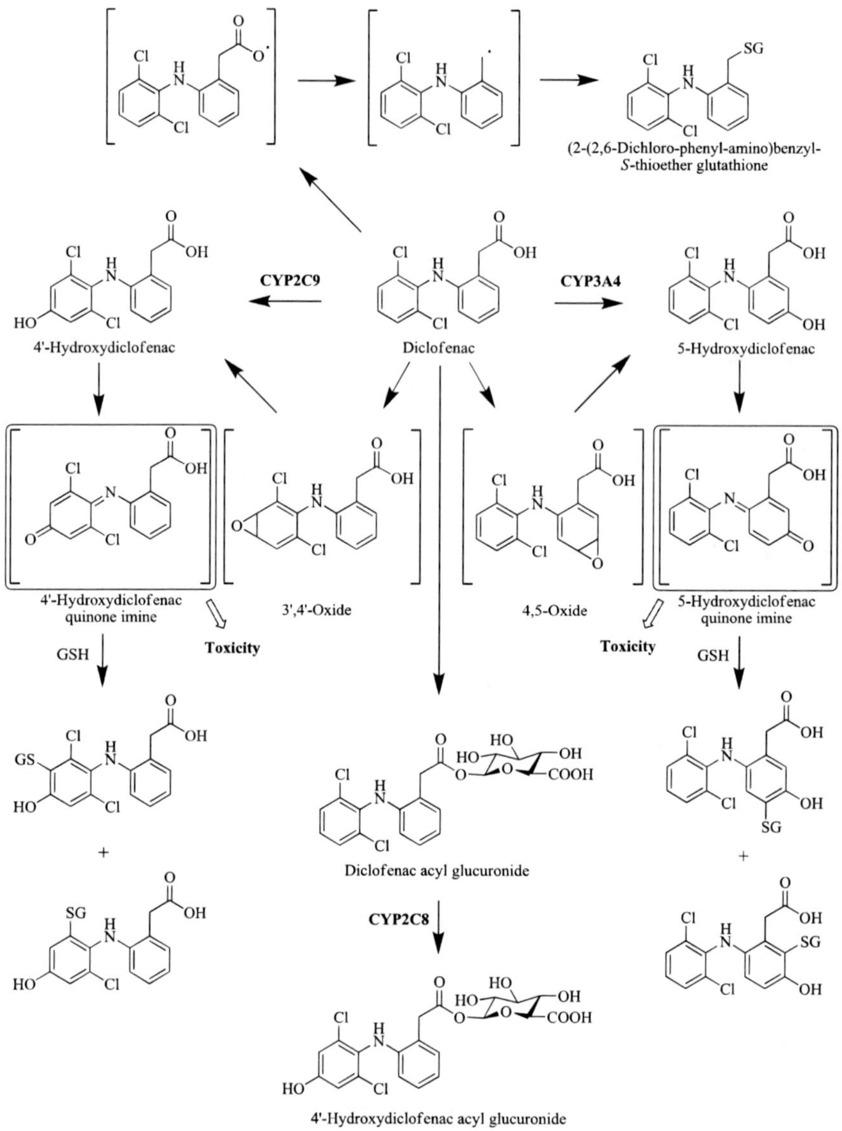

Figure 5. Proposed pathways for the metabolic bioactivation of diclofenac by CYP450 isoenzymes. Diclofenac metabolism can yield potentially toxic reactive intermediates, including quinone-imines, arene oxides and radical species. [62,67,69]

This drug can be responsible for rare but significant cases of serious idiosyncratic hepatotoxicity, [61] and the *in vivo* formation of reactive metabolites has been considered an important factor in its toxicity. [62] Diclofenac is largely metabolized by CYP450 via aromatic hydroxylation mediated by CYP2C9 or CYP3A4 and further oxidation. [63]

These CYP450 isoenzymes exhibit very strict regioselectivities, with CYP2C9 producing 4'-hydroxydiclofenac (major metabolite) and 3'-hydroxydiclofenac and CYP3A4 affording 5-hydroxydiclofenac exclusively [64] Both 4'- and 5-hydroxy metabolites have the potential to be further oxidized into reactive metabolites such as *para*-benzoquinone-imines (Figure 5). These reactive metabolites may originate oxidative stress through redox cycling or covalently binding to non-protein or protein nucleophilic groups [65]. Additionally, arene-oxides, which are reportedly to be the primary intermediates in the metabolization to 4'-hydroxydiclofenac and 5-hydroxydiclofenac, through its covalent binding to a xenobiotic-metabolizing CYP450 isoenzyme in the liver, may induce its irreversible inactivation. Diclofenac acyl glucuronides (Figure 5) are also potential protein-reactive intermediates because of their rather labile ester bond. *In vitro* studies also evidenced that the adducts of 5-hydroxydiclofenac, but not 4'-hydroxydiclofenac, and acyl glucuronide provide an antigenic determinant for immune responses and can result in the onset of a diclofenac-induced idiosyncratic hepatotoxicity [66].

Additional diclofenac metabolites were also identified, namely the GSH-conjugate of the oxidative decarboxylated diclofenac metabolite 2-(2,6-dichlorophenylamino)benzyl-*S*-thioether glutathione (Figure 5), which indicated that the CYP450-mediated toxicity of diclofenac can also be caused by radical species formed during the decarboxylation process. [67] Moreover, since the polymorphic CYP2C9 is involved in the formation of the major metabolite of diclofenac, the relationship between diclofenac metabolic ratios among different CYP2C9 genotypes has been reported, with CYP2C9*3/*3 subjects having a decreased enzyme capacity and, consequently, presenting the highest diclofenac/4'-hydroxydiclofenac ratios [68].

Primaquine

Primaquine (Figure 6), a 8-aminoquinoline derivative, is currently the only drug capable of treating the persistent liver stages of *Plasmodium vivax* and *Plasmodium ovale* responsible for relapsing malaria.

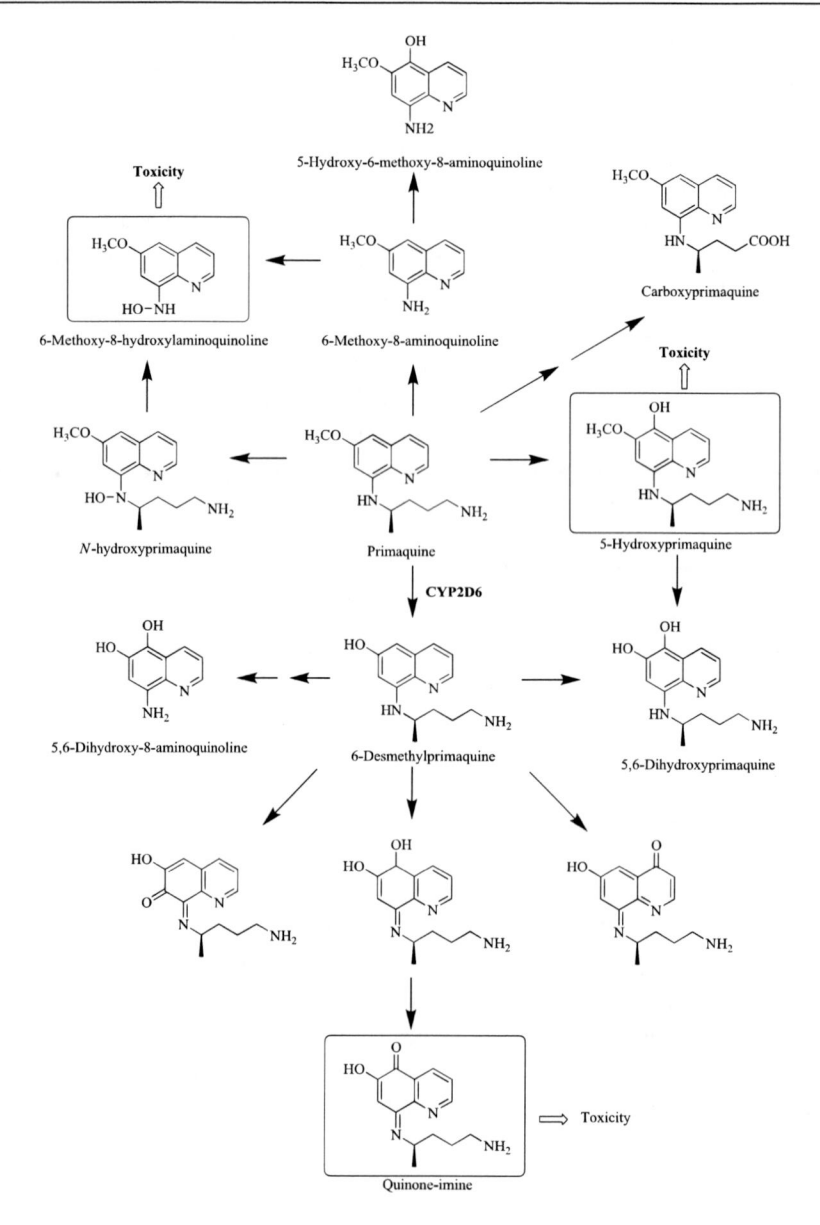

Figure 6. Putative pathways of primaquine metabolism and its toxic metabolites [75, 77].

However, the therapeutic utility of this antimalarial drug has been limited due to hematotoxicity, namely methemoglobinemia and hemolysis, in people with a genetic deficiency of glucose-6-phosphate dehydrogenase. [70, 71]

These toxic effects are believed to be mediated by metabolites, particularly phenolic derivatives which, due to their oxidant nature, can oxidize hemoglobin and generate ROS, thus leading to the depletion of protective thiols. These events finally lead to a dose-related hemolytic anemia [72].

Multiple CYP450 isoforms were reported to contribute to the formation of toxic metabolites and to primaquine hematotoxicity. [73] This compound is rapidly metabolized to carboxyprimaquine, its major metabolite (Figure 6), which is formed by the oxidative deamination of the drug to an aldehyde and its subsequent oxidation. [74] The production of this metabolite may be mediated by monoamine oxidase A (MAO-A) and, to a lesser extent, by CYP2D6. The isoenzyme activity screening and steady-state kinetic data also suggested that CYP3A4, CYP2D6 and CYP2C19 play a role in primaquine metabolism. It was also demonstrated that in this group the isoform CYP2D6 has the highest intrinsic affinity for the metabolism of this drug, particularly in the production of a significant amount of phenolic metabolites. Furthermore, CYP2D6 is associated to a highly polymorphic genetic variability and therefore hemolytic toxicity could be exacerbated in extensive metabolizers. [75] 5-Hydroxyprimaquine also seems to be a putative hematotoxic primaquine metabolite, originating hemolytic anemia *in vivo* (Figure 6). [76] Additionally, structure activity relationship analysis employing primaquine metabolites showed that the metabolites, which could be oxidized to quinones or quinone-imines, generate robust methemoglobin. [72] Primaquine has also been reported to undergo *N*-dealkylation in humans to yield 6-methoxy-8-aminoquinoline, that could be converted to its *N*-hydroxyl analog, 6-methoxy-8-hydroxyaminoquinoline (Figure 6). This compound can also induce methemoglobinemia and hemolytic damage. [74]

Clozapine

Clozapine (Figure 7) is an atypical antipsychotic agent and is generally considered one of the most effective drugs available for the treatment of schizophrenic patients resistant to conventional neuroleptics. It is also useful in acute psychotic attacks, mania, hypomania, and in some other behavioral disorders. Nevertheless, clozapine use seems to be limited by a number of serious adverse events including paralytic ileus, clozapine-induced anticholinergic effects, idiosyncratic hepatotoxicity, cardiac toxicity and agranulocytosis. [78] The biotransformation of this drug by CYP450 enzymes has been proposed as a possible explanation for some adverse events, due to the formation of reactive metabolites. [79] Clozapine undergoes extensive oxidative metabolism of which *N*-demethylation, forming demethylclozapine (the major clozapine metabolite), and *N*-oxidation represent the main metabolic pathways (Figure 7).

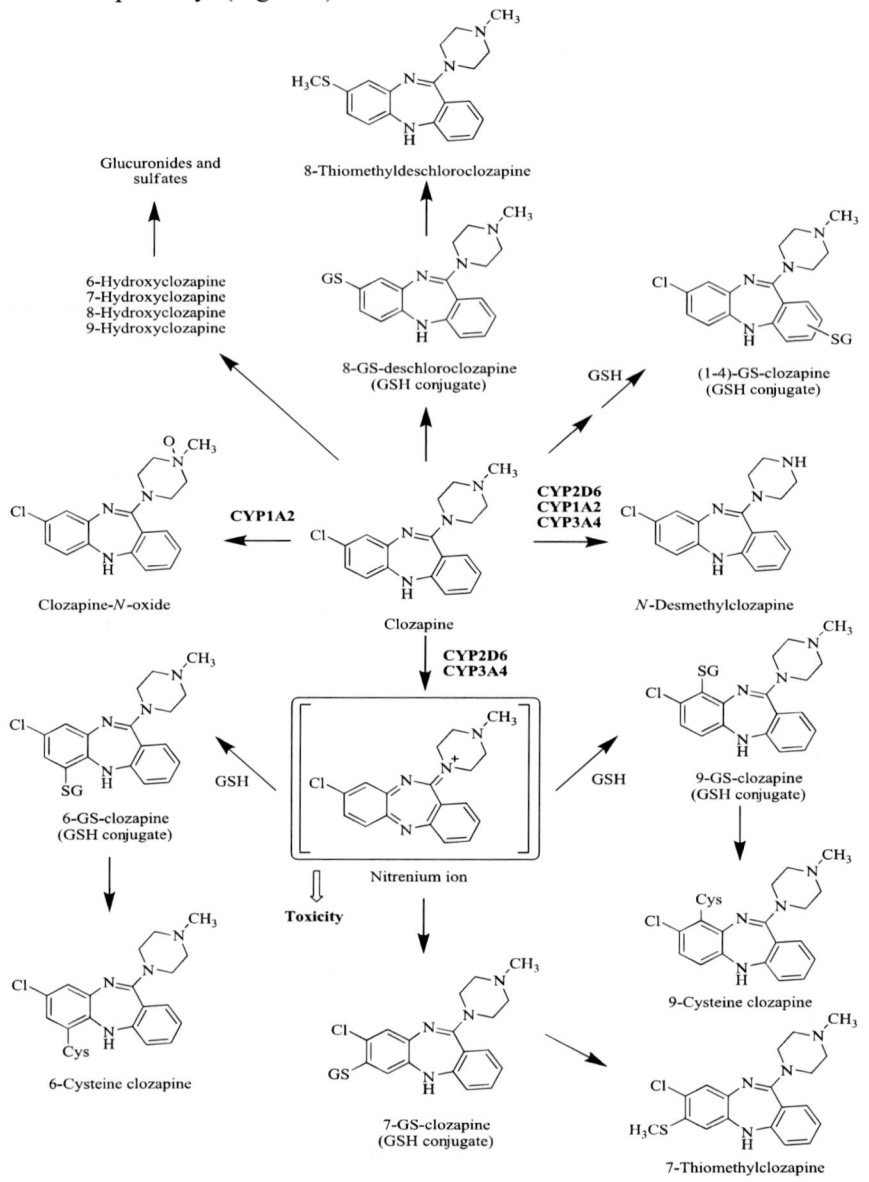

Figure 7. Scheme of clozapine metabolism. [79, 80]

The demethylation reaction is primarily mediated by CYP2D6, CYP1A2 and CYP3A4, whereas CYP1A2 is involved in N-oxidation. [80] Besides these stable metabolites, clozapine is bioactivated by CYP450 enzymes into a reactive nitrenium ion, with CYP3A4 and CYP2D6 showing the highest relevance in this transformation. This ion can then react with GSH to form several GSH conjugates at its 6-, 7-, and 9-positions which can subsequently react with proteins. [79] In the case of peroxidase- and CYP450-mediated bioactivation of clozapine, GSH conjugation mainly occurs at the 6-position of the chlorinated aromatic ring and to a lower extent at the 9-position (Figure 7). In addition, hydroxylation reactions have been demonstrated to occur at the 6-, 7-, 8-, and 9-positions of the chlorinated ring of clozapine (Figure 7). These hydroxylated metabolites are subjected to subsequent glucuronidation and sulfation reactions. [80]

Because CYP1A2 is generally considered the major isoenzyme involved in hepatic clozapine biotransformation to its N-demethylated and N-oxygenated metabolites, [81] interindividual variability in the clearance of this drug could be related to individual CYP1A2 activity. In fact, evidence suggests that CYP1A2*1F/1F genotype may be a risk factor for the lack of therapeutic response to clozapine, particularly in smokers. [82] Additionally, although the CYP2D6 is not the major isoenzyme involved in clozapine metabolism, it has also been a target of research in this context, again due to its polymorphic nature. [83]

Lasofoxifene

Lasofoxifene (Figure 8) is a naphthalene derivative and a third-generation selective estrogen receptor modulator, that was discovered through a synthetic program designed to obtain novel molecules with good oral bioavailability and higher *in vivo* potency. [84]

This new drug completed Phase III development for the prevention and treatment of osteoporosis and was approved in the European Union for this indication in March 2009. From several studies, it was evidenced that lasofoxifene significantly improved bone mineral density and prevented bone loss in postmenopausal women. [85] Its mechanism of action involves the agonism of the estrogen receptor in bone and cholesterol regulation and antagonism of the estrogen receptor in the mammary gland and uterus. [86] Additionally, this nonsteroidal drug has also been studied as a breast cancer growth inhibitor. [87]

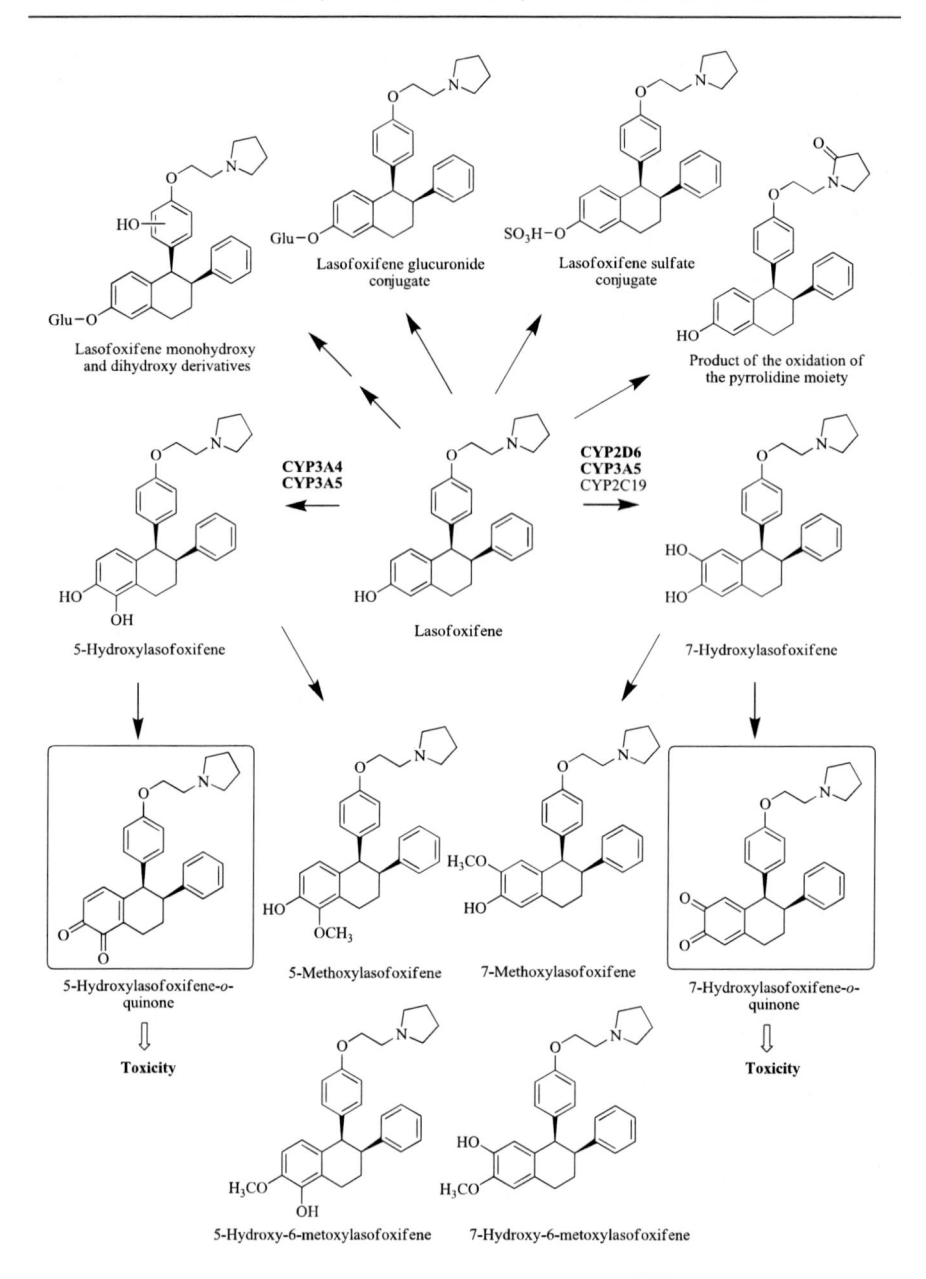

Figure 8. Metabolic pathways of lasoxifene and formation of reactive metabolites. [85, 86]

Lasofoxifene is primarily metabolized in the liver through both oxidative and conjugative pathways. [88] The primary oxidative metabolites are the result of hydroxylations on the tetraline moiety and on the phenyl rings attached to it, and oxidation on the pyrrolidine ring (Figure 8). The tetraline ring hydroxylation can originate two catechol regioisomers: 5-hydroxylasofoxifene and 7-hydroxylasofoxifene (Figure 8).

The first metabolite is formed primarily by CYP3A4 and CYP3A5 activity, whereas 7-hydroxylasofoxifene is originated throught CYP2D6 and CYP3A5 isoforms and, to a much lesser extent, by CYP2C19. Another enzyme involved in this metabolic way is catechol-*O*-methyltransferase which can generate four isomeric monomethyl ethers from these catechol intermediates: 5-methoxylasofoxifene, 5-hydroxy-6-methoxylasofoxifene, 7-methoxylasofoxifene and 7-hydroxy-6-methoxylasofoxifene. [86] The metabolites undergo subsequent glucuronidation and sulfation reactions, forming glucuronide and sulfate conjugates (Figure 8).

In addition, evidence suggested the existence of a further oxidation of 5-hydroxylasofoxifene and 7-hydroxylasofoxifene to the corresponding electrophilic *ortho*-quinones (Figure 8), which can potentially contribute to the known *in vivo* lasofoxifene toxicity. These lasofoxifene-*o*-quinones can be trapped by GSH to originate two mono-GSH and two di-GSH catechol conjugates. However, in the absence of a strong nucleophile like GSH, 7-hydroxylasofoxifene-*o*-quinone can form adducts with DNA. It is believed that CYP1B1 can be involved in lasofoxifene toxicity because these *ortho*-quinones can be generated in tissues such as the breast, ovaries and uterus, where CYP1B1 is primarily expressed. [85]

Nevirapine

Nevirapine (Figure 9) was the first non-nucleoside reverse transcriptase inhibitor approved by the Food and Drug Administration for the treatment of immunodeficiency virus-1 (HIV-1) infection, as part of a combined antiretroviral therapy. [89] This drug is also being used in the prevention of mother-to-child HIV-1 transmission. [90] Although it is very effective, nevirapine can originate serious toxic responses such as hepatotoxicity and skin reactions, [91] which have been associated with its bioactivation.

Nevirapine undergoes Phase I oxidations to several hydroxynevirapine metabolites (2-, 3-, 8-, and 12-hydroxylation) and 4-carboxynevirapine and subsequent Phase II sulfatation to nevirapine-12-sulfate (Figure 9) or glucuronidation. The 2-, 3-, and 8-hydroxylated metabolites can presumably originate *para*-quinone-imine intermediates after further oxidation (Figure 9). [92]

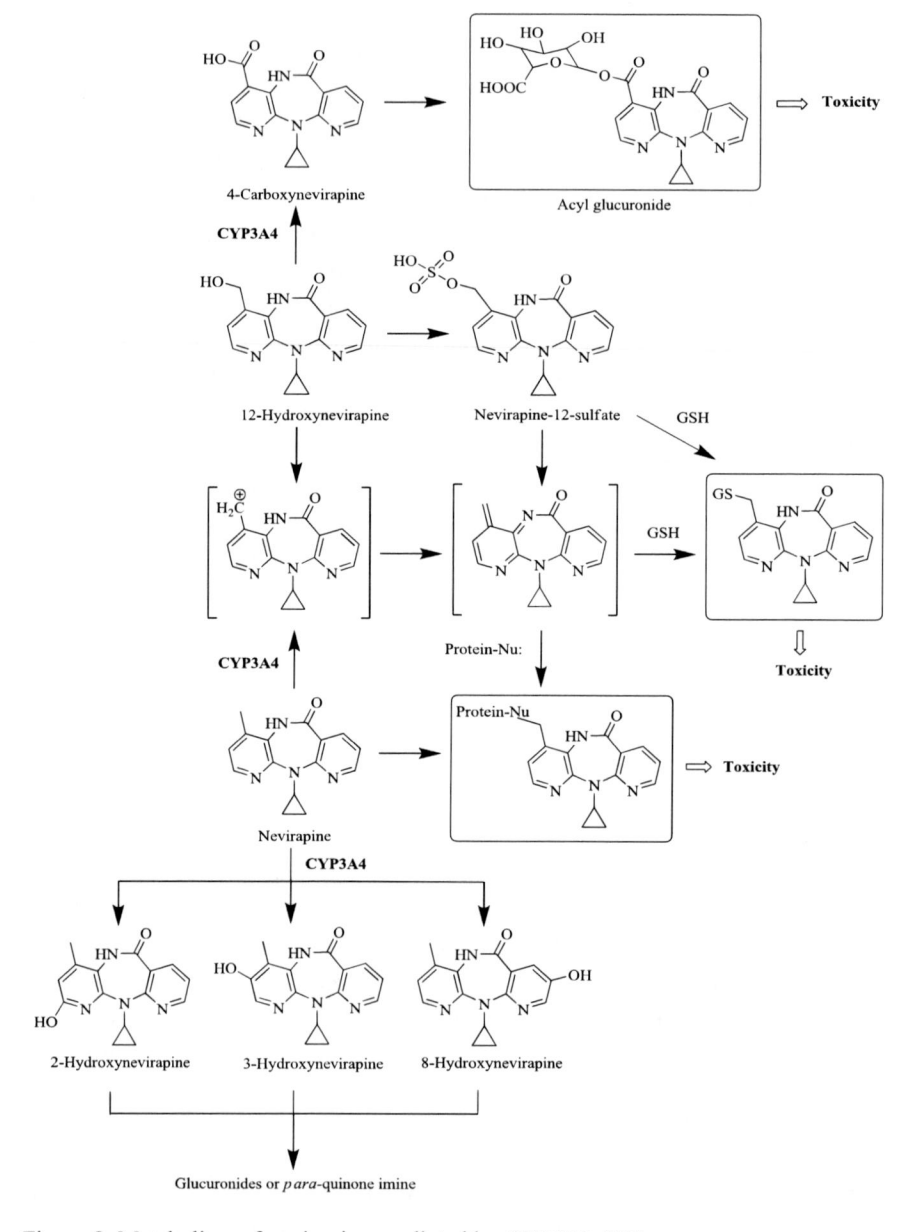

Figure 9. Metabolism of nevirapine mediated by CYP450. [92]

In addition, 4-carboxynevirapine, formed from further oxidation of 12-hydroxynevirapine, can generate a reactive acyl glucuronide, which has the potential to bind to cellular proteins. Several CYP450 isoforms are involved in the metabolism of this drug. Among them, CYP3A4 was identified as the major isoenzyme associated with the formation of 2- and 12-hydroxynevirapine, whereas CYP2B6 was reported to be the predominant isoenzyme that originates 3- and 8-hydroxynevirapine. [92-94]

The metabolite quinone methide is obtained by loss of sulfate from nevirapine-12-sulfate and is a reactive electrophile [92] that forms a GSH adduct derived from the addition reaction of its sulfhydryl nucleophile. [95] Therefore, it is expected that this metabolite also can react directly with biomacromolecules such as DNA, yielding covalent adducts. Recent evidence showed, in fact, that the referred quinone methide may be responsible for nevirapine-induced liver injury, while nevirapine-12-sulfate appears to account for nevirapine-induced skin rash. [96]

Clopidogrel

Clopidogrel (Figure 10) is an antiplatelet agent that irreversibly targets $P2Y_{12}$ adenosine diphosphate receptors on the platelet surface. [97] This molecule is a prodrug that requires hepatic biotransformation to its pharmacologically active thiol intermediates. [98] This activation step is mediated by CYP450 enzymes [99] and by glycoprotein paraoxinase-1. [100] After oral administration, clopidogrel is rapidly absorbed but only a small part is metabolically activated by the CYP450 enzymatic system. The other part (approximately 90%) is converted to an inactive metabolite, clopidogrel carboxylate, by serum and intestinal esterases, mainly human carboxylesterase 1 (hCE1). [101]

Clopidogrel metabolism is reported to occur in two steps [99] (Figure 10). In the first step, its thiophene ring undergoes monooxygenation by hepatic CYP450 isoenzymes, mainly CYP2C19 [101] and to a smaller extent by CYP1A2, CYP2B6 and CYP3A4, [99] producing the intermediate 2-oxo-clopidogrel, a thiolactone. In the second step, this thiolactone is further oxidized through CYP2C19, CYP3A4, CYP2C9 and CYP2B6, leading to the formation of an active thiol metabolite, after suffering a hydrolytic ring-opening and then a GSH-dependent reduction. [99,101,102] The thiol metabolite represents approximately 10% or less of all metabolites. [101]

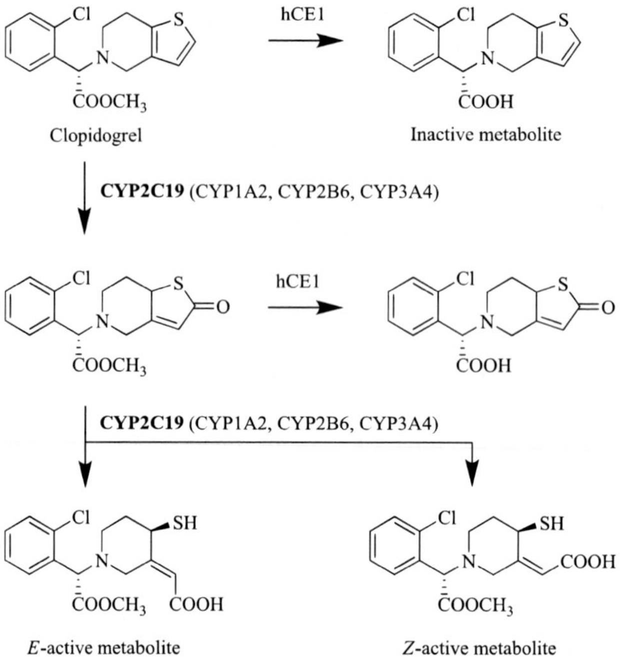

Figure 10. Bioactivation of clopidogrel to inactive and active metabolites. [101]

Recent studies evidenced that clopidogrel metabolism is even more complex and other potentially toxic reactive metabolites can also be originated. These metabolites can form off-target adducts, which can be associated not only with clopidogrel pharmacological action, but also with its toxic effects, such as hepatotoxicity. [99] In addition, Zahno and co-workers [99] showed that incubation of clopidogrel with CYP3A4 is associated with the formation of metabolites that are toxic for hepatocytes. The possible mechanisms explaining clopidogrel toxicity include cellular GSH depletion by reactive metabolites, mitochondrial damage and ROS accumulation which can cause cell death by apoptosis. Therefore, patients with increased production of such metabolites and/or decreased defense systems may be at risk for clopidogrel-associated toxicity.

Methimazole

Methimazole (Figure 11) is widely used in hyperthyroidism. Its main adverse effects are related to skin and gastrointestinal disorders, loss of taste, lupoid-like syndrome, and bone marrow depression with agranulocytosis. Other important adverse event reported for this drug is hepatic injury. [103]

Methimazole is metabolized by CYP450 and flavin monooxygenase to reactive metabolites, and an inadequate detoxification of these metabolites is considered to be responsible for hepatotoxicity. Although the metabolism of this drug is unclear, it is believed that methimazole undergoes epoxidation of the C-4,5 double bond mediated by CYP450 enzymes and, after being hydrolyzed, the resulting epoxide can be decomposed to produce *N*-methylthiourea (Figure 11). Then, flavin-containing monooxygenase (FMO) converts *N*-methylthiourea to its *S*-oxidized metabolites [104] originating reactive sulfenic and sulfinic acids which can produce toxicity. [105] Additionally, Xie and co-workers [105] showed that the CYP2A5 isoenzyme is involved in methimazole toxicity mediated by its metabolic activation.

More recent research evidenced that methimazole causes damages in several tissues/organs (spleen, heart, liver, lung and kidney) because it irreversibly inactivates different peroxidases. [106]

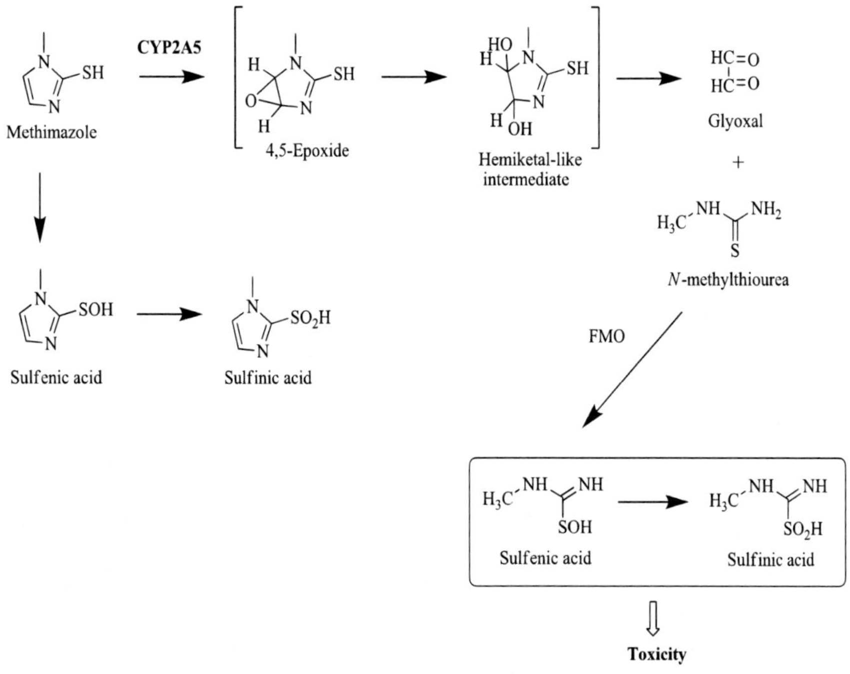

Figure 11. Metabolism of methimazole producing *S*-oxidized metabolites. [104]

Tamoxifen

Tamoxifen (Figure 12) is a selective non-steroidal anti-estrogen drug widely used in the treatment of hormone-dependent breast carcinoma. [107]

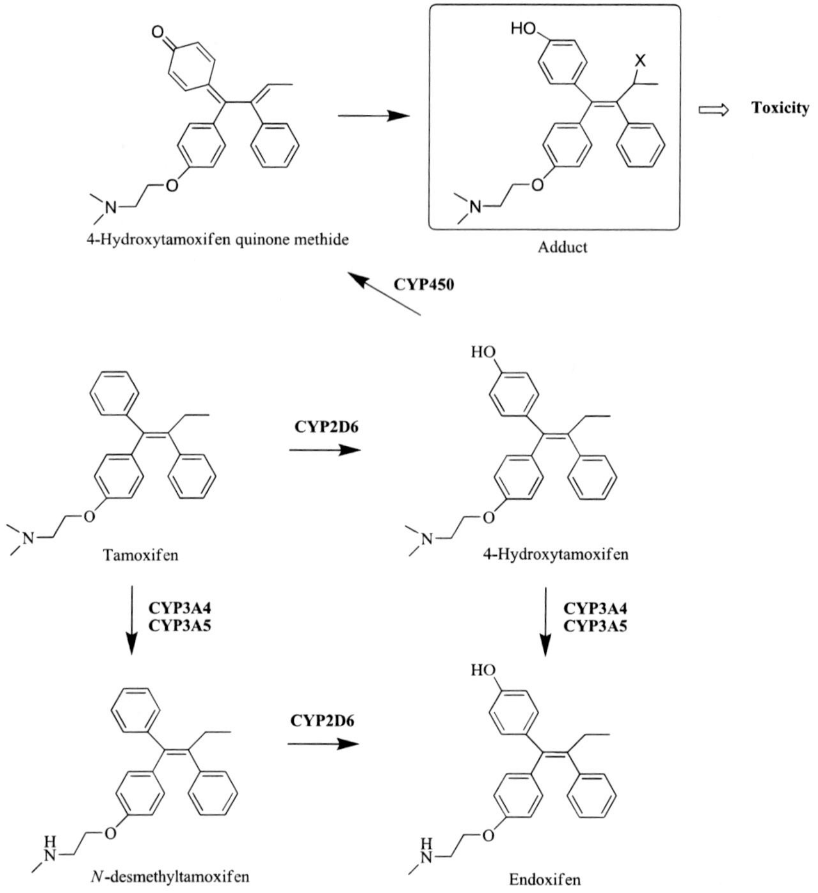

Figure 12. Metabolism of tamoxifen and the role of various CYP450 isoenzymes. The conversion of tamoxifen to N-desmethyltamoxifen is mainly mediated by CYP3A4 and CYP3A5, whereas CYP2D6 plays a major role in the conversion of tamoxifen to 4-hydroxytamoxifen. [111]

This drug is metabolized by enzymes belonging to the CYP450 superfamily and the main isoenzyme involved in its activation is CYP2D6. [108] Its metabolism includes aromatic hydroxylation, demethylation and N-oxidation originating several metabolites that vary in toxicity and estrogenic activity. [109] The isoforms CYP3A4 and CYP3A5 metabolize tamoxifen to N-desmethyltamoxifen, which is transformed by CYP2D6 into the most abundant active metabolite, 4-hydroxy-N-desmethyltamoxifen (endoxifen) (Figure 12). Another important metabolite is 4-hydroxytamoxifen [108,110,111] (Figure 12) which, as endoxifen, presents antiestrogenic effects approximately 100-fold higher than tamoxifen. [108]

Although tamoxifen metabolites are also responsible for its therapeutic effects (mainly endoxifen), they can cause adverse events as well. The side effects of this drug have been primarily ascribed to tamoxifen and N-desmethyltamoxifen plasmatic levels. This metabolite is present in plasma in concentrations 50–135 times higher than 4-hydroxytamoxifen and endoxifen. In addition, several studies suggested that reduced levels of endoxifen are related to a worse prognosis of breast cancer, such as higher recurrence rates and mortality. [108] Indeed, CYP2D6 is a highly polymorphic isoenzyme and thus, lower levels of endoxifen are found in poor metabolizers, possibly due to the reduced CYP2D6 metabolic activity. [111] Shahrokh and co-workers [112] also described that tamoxifen toxicity can be originated by 4-hydroxytamoxifen quinone methide that can form an adduct (Figure 12) with functional proteins.

Terbinafine

Terbinafine (Figure 13) is a synthetic allylamine derivative with antifungal properties and is frequently used in onychomycosis and dermatophytosis treatments. [113]

This drug is extensively metabolized and the main involved reactions are N-demethylation, deamination and oxidation of the tertiary butyl group (Figure 13). [114,115] The CYP450 enzymatic system has an important role in terbinafine metabolism and CYP2C9, CYP1A2 and CYP3A4 are the most important isoenzymes involved in its biotransformation process. More specifically, the N-demethylation is primarily mediated by CYP2C9, CYP2C8 and CYP1A2; the formation of a dihydrodiol is due to the metabolic activity of CYP2C9 and CYP1A2; the deamination is by CYP3A4 and the side chain oxidation is equally originated by CYP1A2, CYP2C8, CYP2C9, as well as CYP2C19. [114] From these reactions, fifteen metabolites were described but the most abundant ones are carboxyterbinafine, desmethylhydroxyterbinafine, and N-desmethylterbinafine. [115,116]

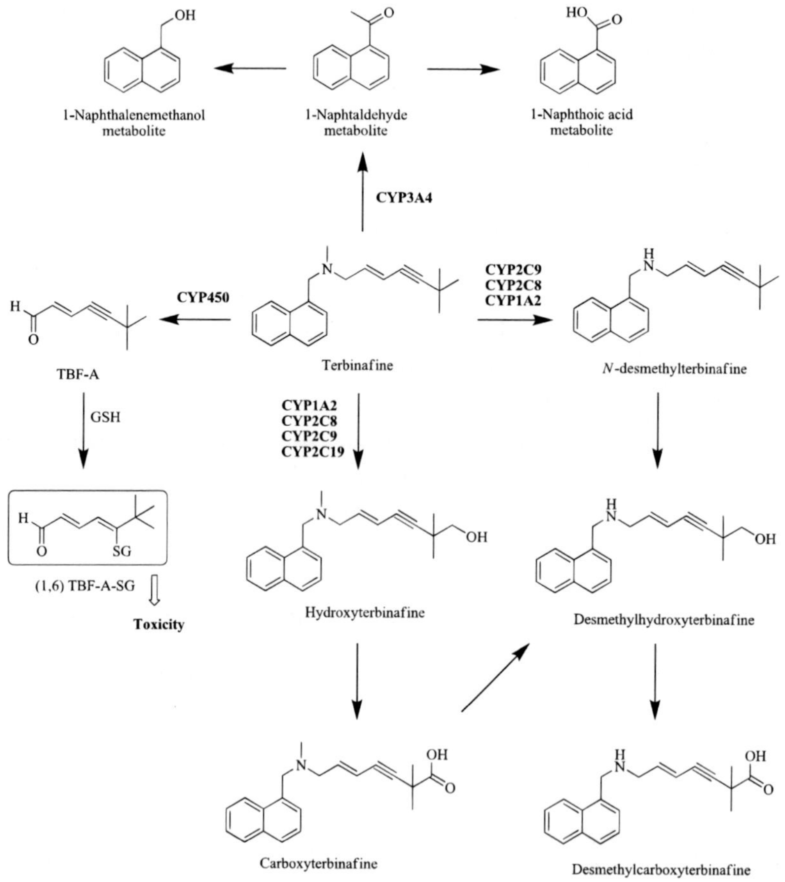

Figure 13. Some metabolic pathways of terbinafine. [114, 115, 117]

Although terbinafine is generally associated with a low toxicity index and few adverse effects, some cases of hepatobiliary dysfunction were linked to its use and were attributed to a reactive metabolite. [116] In this context, Iverson and Uetrecht [117] created a hypothesis to explain the toxicity of this drug (Figure 13). These authors proposed the formation of an allylic aldehyde metabolite, 7,7-dimethylhept-2-ene-4-ynal (TBF-A), by liver enzymes through N-dealkylation, which then suffers conjugation with GSH and is transported across the canalicular membrane of hepatocytes and concentrated in the bile. The mono-GSH conjugate (1,6) TBF-A-SG, which is still reactive, can bind to hepatobiliary proteins leading to direct toxicity. It can also modify canalicular proteins and lead to an immune-mediated reaction causing cholestatic dysfunction. [117]

CONCLUSION

Considering the information presented in this chapter, it can clearly be concluded that CYP450-mediated metabolic reactions are the basis of relevant toxic effects reported to sometimes occur with the use of several common therapeutic drugs. Despite the fact that any part of the human body can be affected, as metabolism is centered in the liver, it is not surprising that the most common form of reported toxicity is hepatotoxicity.

Due to the advances in molecular biology and toxicology, as well as in analytical chemistry, in recent years relevant progress in the knowledge about the metabolic reactions involved in the toxic effects of drugs has been achieved. Several of these recent studies, even involving classic drugs, have highlighted that the metabolic pathways of several drugs are still more complex than it was initially expected, but interesting evolutions in this field are being observed. Thus, the improved knowledge that is being attained should be considered not only in laboratorial studies but also in software algorithms that are being used by medicinal chemists in the design and development of safer drugs. In addition, molecular biology studies are also clarifying the relevance of enzymatic polymorphisms in the therapeutic and toxic effects of drugs, and all these data are of major relevance in the safer use of therapeutic drugs.

ACKNOWLEDGMENTS

Mariana Matias acknowledges the support by Fundação para a Ciência e Tecnologia (Portugal) through the Ph.D. grant (SFRH/BD/85279/2012). The authors also acknowledge Fundação para a Ciência e Tecnologia (Portugal) by the project (PEst-OE/SAU/UI0709/2014).

REFERENCES

[1]   Shakunthala N. New cytochrome P450 mechanisms: implications for understanding molecular basis for drug toxicity at the level of the cytochrome. *Expert Opin Drug Metabol Toxicol* 2010;6:1-15.

[2]   Corsini A, Bortolini M. Drug-induced liver injury: the role of drug metabolism and transport. *J Clin Pharmacol* 2013;53:463-74.

[3]   Sevior DK, Pelkonen O, Ahokas JT. Hepatocytes: the powerhouse of biotransformation. *Int J Biochem Cell Biol* 2012;44:257-61.

[4]   Croom E. Metabolism of xenobiotics of human environments. *Prog Mol Biol Transl Sci* 2012;112:31-88.

[5]   Amacher DE. The effects of cytochrome P450 induction by xenobiotics on endobiotic metabolism in pre-clinical safety studies. *Toxicol Mech Methods* 2010;20:159-66.

[6]   Meunier B, de Visser SP, Shaik S. Mechanism of oxidation reactions catalyzed by cytochrome P450 enzymes. *Chem Rev* 2004;104:3947-80.

[7]   Pelkonen O, Turpeinen M, Hakkola J, Honkakoski P, Hukkanen J, Raunio H. Inhibition and induction of human cytochrome P450 enzymes: current status. *Arch Toxicol* 2008;82:667-715.

[8]   Moktali V, Park J, Fedorova-Abrams ND, et al. Systematic and searchable classification of cytochrome P450 proteins encoded by fungal and oomycete genomes. *BMC Genomics* 2012;13:525.

[9]   Nebert DW, Russell DW. Clinical importance of the cytochromes P450. *Lancet* 2002;360:1155-62.

[10] Omura T, Sato R. The Carbon Monoxide-Binding Pigment of Liver Microsomes: II. Solubilization, purification, and properties. *J Biol Chem* 1964;239:2379-85.

[11] Omura T, Sato R. The Carbon Monoxide-Binding Pigment of Liver Microsomes: I. Evidence for its hemoprotein nature. *J Biol Chem* 1964;239:2370-8.

[12] Yu CA, Gunsalus IC, Katagiri M, Suhara K, Takemori S. Cytochrome P-450_{cam}: I. Crystallization and properties. *J Biol Chem* 1974;249:94-101.

[13] Cosme J, Johnson EF. Engineering microsomal cytochrome P450 2C5 to be a soluble, monomeric enzyme: mutations that alter aggregation, phospholipid dependence of catalysis, and membrane binding. *J Biol Chem* 2000;275:2545-53.

[14] Williams PA, Cosme J, Sridhar V, Johnson EF, McRee DE. Mammalian microsomal cytochrome P450 monooxygenase: structural adaptations for membrane binding and functional diversity. *Mol cell* 2000;5:121-31.

[15] Kemp CA, Maréchal JD, Sutcliffe MJ. Progress in cytochrome P450 active site modeling. *Arch Biochem Biophys* 2005;433:361-8.

[16] Williams PA, Cosme J, Ward A, Angove HC, Matak Vinkovic D, Jhoti H. Crystal structure of human cytochrome P450 2C9 with bound warfarin. *Nature* 2003;424:464-8.

[17] Schoch GA, Yano JK, Wester MR, Griffin KJ, Stout CD, Johnson EF. Structure of human microsomal cytochrome P450 2C8: evidence for a peripheral fatty acid binding site. *J Biol Chem* 2004;279:9497-503.

[18] Williams PA, Cosme J, Vinkovic DM, et al. Crystal structures of human cytochrome P450 3A4 bound to metyrapone and progesterone. *Science* 2004;305:683-6.

[19] Rowland P, Blaney FE, Smyth MG, et al. Crystal structure of human cytochrome P450 2D6. *J Biol Chem* 2006;281:7614-22.

[20] Sangar MC, Bansal S, Avadhani NG. Bimodal targeting of microsomal cytochrome P450s to mitochondria: implications in drug metabolism and toxicity. *Expert Opin Drug Metabol Toxicol* 2010;6:1231-51.

[21] Cribb AE, Peyrou M, Muruganandan S, Schneider L. The endoplasmic reticulum in xenobiotic toxicity. *Drug Metabol Rev* 2005;37:405-42.

[22] Zanger UM, Schwab M. Cytochrome P450 enzymes in drug metabolism: regulation of gene expression, enzyme activities, and impact of genetic variation. *Pharmacol Ther* 2013;138:103-41.

[23] Wang JF, Chou KC. Molecular modeling of cytochrome P450 and drug metabolism. *Curr Drug Metabol* 2010;11:342-6.

[24] McKinnon RA, Sorich MJ, Ward MB. Cytochrome P450 Part1: Multiplicity and Function. *J Pharm Pract Res* 2008;38:55-7.

[25] Guengerich FP. Cytochrome P450s and other enzymes in drug metabolism and toxicity. *AAPS J* 2006;8:E101-11.

[26] Madanayake TW, Lindquist IE, Devitt NP, Mudge J, Rowland AM. A transcriptomic approach to elucidate the physiological significance of human cytochrome P450 2S1 in bronchial epithelial cells. *BMC Genomics* 2013;14:833.

[27] Mishra NK. Computational modeling of P450s for toxicity prediction. *Expert Opin Drug Metabol Toxicol* 2011;7:1211-31.

[28] Liu YT, Hao HP, Liu CX, Wang GJ, Xie HG. Drugs as CYP3A probes, inducers, and inhibitors. *Drug Metabol Rev* 2007;39:699-721.

[29] Brown CM, Reisfeld B, Mayeno AN. Cytochromes P450: a structure-based summary of biotransformations using representative substrates. *Drug Metabol Rev* 2008;40:1-100.

[30] Hart SN, Zhong XB. P450 oxidoreductase: genetic polymorphisms and implications for drug metabolism and toxicity. *Expert Opin Drug Metabol Toxicol* 2008;4:439-52.

[31] Johansson I, Ingelman-Sundberg M. Genetic polymorphism and toxicology - with emphasis on cytochrome P450. *Toxicol Sci* 2011;120:1-13.

[32] Masubuchi Y, Horie T. Toxicological significance of mechanism-based inactivation of cytochrome P450 enzymes by drugs. *Crit Rev Toxicol* 2007;37:389-412.

[33] Masic LP. Role of cyclic tertiary amine bioactivation to reactive iminium species: structure toxicity relationship. *Curr Drug Metabol* 2011;12:35-50.

[34] Guengerich FP, MacDonald JS. Applying mechanisms of chemical toxicity to predict drug safety. *Chem Res Toxicol* 2007;20:344-69.

[35] Edwards PJ, Sturino C. Managing the liabilities arising from structural alerts: a safe philosophy for medicinal chemists. *Curr Med Chem* 2011;18:3116-35.

[36] Walsh JS, Miwa GT. Bioactivation of drugs: risk and drug design. Annu *Rev Pharmacol Toxicol* 2011;51:145-67.

[37] Williams DP. Toxicophores: investigations in drug safety. *Toxicology* 2006;226:1-11.

[38] Kalgutkar AS, Didiuk MT. Structural alerts, reactive metabolites, and protein covalent binding: how reliable are these attributes as predictors of drug toxicity? *Chem Biodivers* 2009;6:2115-37.

[39] Kalgutkar AS, Gardner I, Obach RS, et al. A comprehensive listing of bioactivation pathways of organic functional groups. *Curr Drug Metabol* 2005;6:161-225.

[40] Zhou S, Chan E, Duan W, Huang M, Chen YZ. Drug bioactivation, covalent binding to target proteins and toxicity relevance. *Drug Metabol* Rev 2005;37:41-213.

[41] Guengerich FP. Mechanisms of cytochrome P450 substrate oxidation: minireview. *J Biochem Mol Toxicol* 2007;21:163-8.

[42] Deavall DG, Martin EA, Horner JM, Roberts R. Drug-induced oxidative stress and toxicity. *J Toxicol* 2012;2012:645460.

[43] Ma Q. Role of Nrf2 in oxidative stress and toxicity. *Annu Rev Pharmacol Toxicol* 2013;53:401-26.

[44] Guengerich FP. Common and uncommon cytochrome P450 reactions related to metabolism and chemical toxicity. *Chem Res Toxicol* 2001;14:611-50.

[45] Boelsterli UA, Ho HK, Zhou S, Leow KY. Bioactivation and hepatotoxicity of nitroaromatic drugs. *Curr Drug Metabol* 2006;7:715-27.

[46] Williams DP, Park BK. Idiosyncratic toxicity: the role of toxicophores and bioactivation. *Drug Discov Today* 2003;8:1044-50.

[47] Jaeschke H, McGill MR, Ramachandran A. Oxidant stress, mitochondria, and cell death mechanisms in drug-induced liver injury: lessons learned from acetaminophen hepatotoxicity. *Drug Metabol Rev* 2012;44:88-106.

[48] Zhao L, Pickering G. Paracetamol metabolism and related genetic differences. *Drug Metabol Rev* 2011;43:41-52.

[49] Tonoli D, Varesio E, Hopfgartner G. Quantification of acetaminophen and two of its metabolites in human plasma by ultra-high performance liquid chromatography-low and high resolution tandem mass spectrometry. *J Chromatogr B* 2012;904:42-50.

[50] James LP, Chiew A, Abdel-Rahman SM, et al. Acetaminophen protein adduct formation following low-dose acetaminophen exposure: comparison of immediate-release vs extended-release formulations. *Eur J Clin Pharmacol* 2013;69:851-7.

[51] Abdelmegeed MA, Moon KH, Chen C, Gonzalez FJ, Song BJ. Role of cytochrome P450 2E1 in protein nitration and ubiquitin-mediated degradation during acetaminophen toxicity. *Biochem Pharmacol* 2010;79:57-66.

[52] Hadi M, Dragovic S, van Swelm R, et al. AMAP, the alleged non-toxic isomer of acetaminophen, is toxic in rat and human liver. *Arch Toxicol* 2013;87:155-65.

[53] Graham GG, Davies MJ, Day RO, Mohamudally A, Scott KF. The modern pharmacology of paracetamol: therapeutic actions, mechanism of action, metabolism, toxicity and recent pharmacological findings. *Inflammopharmacol* 2013;21:201-32.

[54] An JH, Lee HJ, Jung BH. Quantitative analysis of acetaminophen and its six metabolites in rat plasma using liquid chromatography/tandem mass spectrometry. *Biomed Chromatogr* 2012;26:1596-604.

[55] Ho KM, Power BM. Benefits and risks of furosemide in acute kidney injury. Anaesthesia 2010;65:283-93.

[56] Peterson LA. Reactive metabolites in the biotransformation of molecules containing a furan ring. *Chem Res Toxicol* 2013;26:6-25.

[57] Yang KH, Choi YH, Lee U, Lee JH, Lee MG. Effects of cytochrome P450 inducers and inhibitors on the pharmacokinetics of intravenous furosemide in rats: involvement of CYP2C11, 2E1, 3A1 and 3A2 in furosemide metabolism. *J Pharm Pharmacol* 2009;61:47-54.

[58] Williams DP, Antoine DJ, Butler PJ, et al. The metabolism and toxicity of furosemide in the Wistar rat and CD-1 mouse: a chemical and

biochemical definition of the toxicophore. *J Pharmacol Exp Ther* 2007;322:1208-20.

[59] Mondal SC, Tripathi DN, Vikram A, Ramarao P, Jena GB. Furosemide-induced genotoxicity and cytotoxicity in the hepatocytes, but weak genotoxicity in the bone marrow cells of mice. *Fundam Clin Pharmacol* 2012;26:383-92.

[60] Gan TJ. Diclofenac: an update on its mechanism of action and safety profile. *Curr Med Res Opin* 2010;26:1715-31.

[61] Mueller D, Muller-Vieira U, Biemel KM, Tascher G, Nussler AK, Noor F. Biotransformation of diclofenac and effects on the metabolome of primary human hepatocytes upon repeated dose exposure. *Eur J Pharm Sci* 2012;45:716-24.

[62] Boerma JS, Dragovic S, Vermeulen NP, Commandeur JN. Mass spectrometric characterization of protein adducts of multiple P450-dependent reactive intermediates of diclofenac to human glutathione-S-transferase P1-1. *Chem Res Toxicol* 2012;25:2532-41.

[63] Lauer B, Tuschl G, Kling M, Mueller SO. Species-specific toxicity of diclofenac and troglitazone in primary human and rat hepatocytes. *Chem Biol Interact* 2009;179:17-24.

[64] Lonsdale R, Houghton KT, Zurek J, et al. Quantum mechanics/molecular mechanics modeling of regioselectivity of drug metabolism in cytochrome P450 2C9. *J Am Chem Soc* 2013;135:8001-15.

[65] Ramm S, Mally A. Role of drug-independent stress factors in liver injury associated with diclofenac intake. *Toxicology* 2013;312:83-96.

[66] Kishida T, Onozato T, Kanazawa T, Tanaka S, Kuroda J. Increase in covalent binding of 5-hydroxydiclofenac to hepatic tissues in rats co-treated with lipopolysaccharide and diclofenac: involvement in the onset of diclofenac-induced idiosyncratic hepatotoxicity. *J Toxicol Sci* 2012;37:1143-56.

[67] van Leeuwen JS, Vredenburg G, Dragovic S, Tjong TF, Vos JC, Vermeulen NP. Metabolism related toxicity of diclofenac in yeast as model system. *Toxicol Lett* 2011;200:162-8.

[68] Dorado P, Berecz R, Cáceres MC, LLerena A. Analysis of diclofenac and its metabolites by high-performance liquid chromatography: relevance of CYP2C9 genotypes in diclofenac urinary metabolic ratios. *J Chromatogr B* 2003;789:437-42.

[69] Boelsterli UA. Diclofenac-induced liver injury: a paradigm of idiosyncratic drug toxicity. Toxicol Appl Pharmacol 2003;192:307-22.

[70] Baird JK, Hoffman SL. Primaquine therapy for malaria. *Clin Infect Dis* 2004;39:1336-45.

[71] Fernando D, Rodrigo C, Rajapakse S. Primaquine in vivax malaria: an update and review on management issues. *Malar J* 2011;10:351.

[72] Ganesan S, Chaurasiya ND, Sahu R, Walker LA, Tekwani BL. Understanding the mechanisms for metabolism-linked hemolytic toxicity of primaquine against glucose 6-phosphate dehydrogenase deficient human erythrocytes: evaluation of eryptotic pathway. *Toxicology* 2012;294:54-60.

[73] Ganesan S, Tekwani BL, Sahu R, Tripathi LM, Walker LA. Cytochrome P_{450}-dependent toxic effects of primaquine on human erythrocytes. *Toxicol Appl Pharmacol* 2009;241:14-22.

[74] Avula B, Tekwani BL, Chaurasiya ND, et al. Profiling primaquine metabolites in primary human hepatocytes using UHPLC-QTOF-MS with ^{13}C stable isotope labeling. *J Mass Spectrom* 2013;48:276-85.

[75] Pybus BS, Sousa JC, Jin X, et al. CYP450 phenotyping and accurate mass identification of metabolites of the 8-aminoquinoline, anti-malarial drug primaquine. *Malar J* 2012;11:259.

[76] Bowman ZS, Jollow DJ, McMillan DC. Primaquine-induced hemolytic anemia: role of splenic macrophages in the fate of 5-hydroxyprimaquine-treated rat erythrocytes. *J Pharmacol Exp Ther* 2005;315:980-6.

[77] Bolchoz LJ, Budinsky RA, McMillan DC, Jollow DJ. Primaquine-induced hemolytic anemia: formation and hemotoxicity of the arylhydroxylamine metabolite 6-methoxy-8-hydroxylaminoquinoline. *J Pharmacol Exp Ther* 2001;297:509-15.

[78] Fitzsimons J, Berk M, Lambert T, Bourin M, Dodd S. A review of clozapine safety. *Expert Opin Drug Saf* 2005;4:731-44.

[79] Vredenburg G, Vassell KP, Commandeur JN, Vermeulen NP, Vos JC. Reconstitution of the interplay between cytochrome P450 and human glutathione S-transferases in clozapine metabolism in yeast. *Toxicol Lett* 2013;222:247-56.

[80] Dragovic S, Boerma JS, van Bergen L, Vermeulen NP, Commandeur JN. Role of human glutathione *S*-transferases in the inactivation of reactive metabolites of clozapine. *Chem Res Toxicol* 2010;23:1467-76.

[81] Zhang WV, D'Esposito F, Edwards RJ, Ramzan I, Murray M. Interindividual variation in relative CYP1A2/3A4 phenotype influences susceptibility of clozapine oxidation to cytochrome P450-specific

inhibition in human hepatic microsomes. *Drug Metabol Dispos* 2008;36:2547-55.

[82] Balibey H, Basoglu C, Lundgren S, et al. CYP1A2*1F Polymorphism Decreases Clinical Response to Clozapine in Patients with Schizophrenia. *Bull Clin Psychopharmacol* 2011;21:93-9.

[83] Gregoor JG, van der Weide K, van der Weide J, van Megen HJ, Egberts AC, Heerdink ER. The association between CYP2D6 genotype and switching antipsychotic medication to clozapine. *Eur J Clin Pharmacol* 2013;69:1927-32.

[84] Gennari L, Merlotti D, Nuti R. Selective estrogen receptor modulator (SERM) for the treatment of osteoporosis in postmenopausal women: focus on lasofoxifene. *Clin Interv Aging* 2010;5:19-29.

[85] Michalsen BT, Gherezghiher TB, Choi J, et al. Selective estrogen receptor modulator (SERM) lasofoxifene forms reactive quinones similar to estradiol. *Chem Res Toxicol* 2012;25:1472-83.

[86] Prakash C, Johnson KA, Gardner MJ. Disposition of lasofoxifene, a next-generation selective estrogen receptor modulator, in healthy male subjects. *Drug Metabol Dispos* 2008;36:1218-26.

[87] LaCroix AZ, Powles T, Osborne CK, et al. Breast cancer incidence in the randomized PEARL trial of lasofoxifene in postmenopausal osteoporotic women. *J Natl Cancer Inst* 2010;102:1706-15.

[88] Gennari L, Merlotti D, Martini G, Nuti R. Lasofoxifene: a third-generation selective estrogen receptor modulator for the prevention and treatment of osteoporosis. *Expert Opin Investig Drugs* 2006;15:1091-103.

[89] de Bethune MP. Non-nucleoside reverse transcriptase inhibitors (NNRTIs), their discovery, development, and use in the treatment of HIV-1 infection: a review of the last 20 years (1989-2009). *Antiviral Res* 2010;85:75-90.

[90] Dellamonica P, Di Perri G, Garraffo R. NNRTIs: pharmacological data. *Med Mal Infect* 2012;42:287-95.

[91] Dellamonica P, Di Perri G, Garraffo R. NNRTIs: future prospects. *Med Mal Infect* 2012;42:296-300.

[92] Wen B, Chen Y, Fitch WL. Metabolic activation of nevirapine in human liver microsomes: dehydrogenation and inactivation of cytochrome P450 3A4. *Drug Metabol Dispos* 2009;37:1557-62.

[93] Fang JL, Beland FA. Differential responses of human hepatocytes to the non-nucleoside HIV-1 reverse transcriptase inhibitor nevirapine. *J Toxicol Sci* 2013;38:741-52.

[94] Fan-Havard P, Liu Z, Chou M, et al. Pharmacokinetics of phase I nevirapine metabolites following a single dose and at steady state. *Antimicrob Agents Chemother* 2013;57:2154-60.

[95] Srivastava A, Lian LY, Maggs JL, et al. Quantifying the metabolic activation of nevirapine in patients by integrated applications of NMR and mass spectrometries. *Drug Metabol Dispos* 2010;38:122-32.

[96] Antunes AM, Wolf B, Oliveira MC, Beland FA, Marques MM. 2'-Deoxythymidine adducts from the anti-HIV drug nevirapine. *Molecules* 2013;18:4955-71.

[97] Berger JS. Aspirin, clopidogrel, and ticagrelor in acute coronary syndromes. Am J Cardiol 2013;112:737-45.

[98] Kleiman NS. Searching for the ceiling: new reflections on the disposition and metabolism of clopidogrel. *JACC Cardiovasc Interv* 2008;1:628-30.

[99] Zahno A, Bouitbir J, Maseneni S, Lindinger PW, Brecht K, Krahenbuhl S. Hepatocellular toxicity of clopidogrel: mechanisms and risk factors. *Free Radic Biol Med* 2013;65:208-16.

[100] Namazi S, Kojuri J, Khalili A, Azarpira N. The impact of genetic polymorphisms of *P2Y12*, *CYP3A5* and *CYP2C19* on clopidogrel response variability in Iranian patients. *Biochem Pharmacol* 2012;83:903-8.

[101] Elsinghorst PW. Quantitative determination of clopidogrel and its metabolites in biological samples: a mini-review. *J Chromatogr B* 2013;917-918:48-52.

[102] Bates ER, Lau WC, Angiolillo DJ. Clopidogrel-drug interactions. *J Am Coll Cardiol* 2011;57:1251-63.

[103] Franklyn JA, Boelaert K. Thyrotoxicosis. *Lancet* 2012;379:1155-66.

[104] Mizutani T, Yoshida K, Murakami M, Shirai M, Kawazoe S. Evidence for the involvement of *N*-methylthiourea, a ring cleavage metabolite, in the hepatotoxicity of methimazole in glutathione-depleted mice: structure-toxicity and metabolic studies. *Chem Res Toxicol* 2000;13:170-6.

[105] Xie F, Zhou X, Genter MB, Behr M, Gu J, Ding X. The tissue-specific toxicity of methimazole in the mouse olfactory mucosa is partly mediated through target-tissue metabolic activation by CYP2A5. *Drug Metabol Dispos* 2011;39:947-51.

[106] Cano-Europa E, Blas-Valdivia V, Franco-Colin M, Gallardo-Casas CA, Ortiz-Butron R. Methimazole-induced hypothyroidism causes cellular

damage in the spleen, heart, liver, lung and kidney. *Acta Histochem* 2011;113:1-5.

[107] den Hollander P, Savage MI, Brown PH. Targeted therapy for breast cancer prevention. *Front Oncol* 2013;3:250.

[108] Antunes MV, Rosa DD, Viana TS, Andreolla H, Fontanive TO, Linden R. Sensitive HPLC-PDA determination of tamoxifen and its metabolites N-desmethyltamoxifen, 4-hydroxytamoxifen and endoxifen in human plasma. *J Pharm Biomed Anal* 2013;76:13-20.

[109] Teunissen SF, Rosing H, Seoane MD, et al. Investigational study of tamoxifen phase I metabolites using chromatographic and spectroscopic analytical techniques. *J Pharm Biomed Anal* 2011;55:518-26.

[110] Irvin WJ, Jr., Walko CM, Weck KE, et al. Genotype-guided tamoxifen dosing increases active metabolite exposure in women with reduced CYP2D6 metabolism: a multicenter study. *J Clin Oncol* 2011;29:3232-9.

[111] Singh MS, Francis PA, Michael M. Tamoxifen, cytochrome P450 genes and breast cancer clinical outcomes. *Breast* 2011;20:111-8.

[112] Shahrokh K, Cheatham III TE, Yost GS. Conformational dynamics of CYP3A4 demonstrate the important role of Arg212 coupled with the opening of ingress, egress and solvent channels to dehydrogenation of 4-hydroxy-tamoxifen. *Biochim Biophys Acta* 2012;1820:1605-17.

[113] Krishnan-Natesan S. Terbinafine: a pharmacological and clinical review. *Expert Opin Pharmacother* 2009;10:2723-33.

[114] Vickers AE, Sinclair JR, Zollinger M, et al. Multiple cytochrome P-450s involved in the metabolism of terbinafine suggest a limited potential for drug-drug interactions. *Drug Metabol Dispos* 1999;27:1029-38.

[115] Zehender H, Denouel J, Roy M, Le Saux L, Schaub P. Simultaneous determination of terbinafine (Lamisil) and five metabolites in human plasma and urine by high-performance liquid chromatography using on-line solid-phase extraction. *J Chromatogr B* 1995;664:347-55.

[116] Keller KA. Therapeutic review: terbinafine. *J Exotic Pet Med* 2012;21:181-5.

[117] Iverson SL, Uetrecht JP. Identification of a reactive metabolite of terbinafine: insights into terbinafine-induced hepatotoxicity. *Chem Res Toxicol* 2001;14:175-81.

In: Cytochrome P450 Enzymes
Editor: Jian Wu

ISBN: 978-1-61942-209-4
© 2014 Nova Science Publishers, Inc.

Chapter 3

EVALUATION OF CYTOCHROME P450 ENZYMES IN DRUG INTERACTIONS

Xin Wang[*]

Shanghai Key Laboratory of Regulatory Biology,
Institute of Biomedical Sciences and School of Life Sciences,
East China Normal University, Shanghai, China

ABSTRACT

Cytochrome P450 (CYP) enzyme is a super family of the Phase I enzymes in the biotransformation of xenobiotics. Most drugs undergo deactivation by CYP, either directly or by facilitated excretion from the body. In addition, CYP plays a primary role in drug interactions that can result in drug toxicity, reduced pharmacological effect, and adverse drug reactions. At the same time, many drugs also have the ability to affect CYP expression and activity that may have the potential to interfere with drug metabolism. Although the clinical importance of drug interactions depends on many factors associated with the particular drug and patient, recognizing whether the drug involved acts as a CYP enzyme substrate, inducer, or inhibitor can prevent clinically significant interactions from occurring. This chapter briefly reviews the mechanisms for drug interactions, problems in monitoring drug interactions, methods to assess drug interactions, and the role of CYP enzymes in drug interactions. A better understanding of interactions of drugs with CYP will help the

[*] E-mail: usxinwang@gmail.com, xwang@bio.ecnu.edu.cn, Tel: +86 -21-2420 6564, Fax: +86-21-5434 4922.

regulation of the use of drugs, avoid co-administration or anticipating potential problems, and adjust a patient's drug dose early in order to provide optimal response with minimal adverse effects.

Keywords: Cytochrome P450 (CYP), drug interactions, metabolism, toxicity

INTRODUCTION

Drug-drug interactions are a common issue during drug treatment and often give rise to a large number of hospital admissions in the world. Moreover, there is a belief among the general public that herbs are safe health supplements. If herbs, however, are used in combination with therapeutic drugs, it may also raise the potential of pharmacokinetic and/or pharmacodynamic herb-drug interactions. In fact, the drug-drug and/or herb-drug interactions may lead to serious clinical consequences. For example, there have been reports of herb-drug interactions between warfarin and herbs such as St. John's wort [1, 2], Danshen [3], Dong quai [4], ginseng [5], and ginkgo [6, 7].

Combined use of drugs and/or herbs may mimic, increase or reduce the effects of either component, which may result in clinically important drug interactions. It is possible that one substance may alter the bioavailability and the clinical effectiveness of another substance when given concurrently. The consequence can be either an exaggeration of pharmacological or toxic effect or a diminished efficacy of drug treatment. This interaction, for instance, will endanger patients' condition as demonstrated by the published reports in which enhanced anticoagulation and bleeding occurred when patients on warfarin therapy consumed Danshen [3, 6, 8, 9]. Both pharmacokinetic and pharmacodynamic mechanisms have been considered to play the important role in these interactions, although the underlying mechanisms for the altered drug effects and concentrations by concomitant medicines are yet to be determined.

The clinical importance of drug interactions depends on many factors associated with the particular drug, herb and patient. Furthermore, there are also some problems in monitoring drug interactions. Therefore, the potential interactions when concomitantly used with drugs and/or herbs should be appropriately labeled to alert consumers.

Cytochrome P450 (CYP) monooxygenases belong to a superfamily of haem-proteins that play an important role in catalyzing the oxidation of many endogenous and exogenous substances. The activity of CYP enzymes has been reported to vary between individuals up to 50-fold for some index metabolic reactions [10].

Several factors affect P450 enzyme activity, including genetic polymorphisms, age, gender, disease states, and environmental influences such as smoking or exposure to chemicals [11]. CYP enzymes are primarily located in the liver, and are widely distributed in other tissues, such as intestine, lung, kidney, and brain [11]. Drug interactions usually occur as CYP has broad substrate specificity and overlapping substrate selectivity [12]. In addition, the activity of CYP can be altered by enzyme induction or inhibition, thus affecting the disposition and pharmacological action of co-administered drugs. Among the various CYP isoenzymes, CYP1A2, 2C9, 2C19, 2D6 and 3A4 are major isoforms responsible for more than 90% metabolism of drugs in humans [13, 14].

This chapter briefly reviews the mechanisms for drug interactions, problems in monitoring drug interactions, methods to assessing drug interactions, and the role of CYP enzyme in drug interactions. A better understanding of interactions of drugs with CYP will help the regulation of the use of drugs, avoid co-administration or anticipating potential problems, and adjust a patient's drug dose early in order to provide an optimal response with minimal adverse effects.

MECHANISMS FOR DRUG INTERACTIONS

The mechanisms for drug interactions can be divided into two types: pharmacokinetic and pharmacodynamic interactions [15]. Pharmacokinetic interactions may result from alteration of absorption, distribution, metabolism or elimination of a conventional drug by an herbal product or other dietary supplements [16]. Pharmacodynamic interactions may occur when drugs and/or constituents of herbal products produce synergistic or antagonistic activity in relation to a conventional drug.

As a result, concentration-dependent activity of a therapeutic molecule is altered at the site of action at the drug-receptor level. Although many herb-drug interactions are likely to be neglected in nature, it is important to realize that some interactions may have an exaggeration of pharmacological or toxic effect on drug therapy [17].

Pharmacokinetic Interactions

Pharmacokinetic interactions occur when one drug may alter the rate or extent of absorption, distribution, metabolism or elimination on a combination of these processes of another drug. The interaction frequently causes marked shifts in serum drug levels and may alter clinical responses. It is most commonly measured by a change in one or more kinetic parameters such as maximum serum concentration (C_{max}), area under the concentration-time curve (AUC), half-life, and total amount of drug excreted in urine, etc. [18]. Many commonly used herbs have been shown to modulate the plasma clearance of important therapeutic drugs. For example, clinical studies have shown that St John's wort reduced the AUC of cyclosporin [19, 20], amitriptyline [21], digoxin [22], indinavir [23], nevirapine [24], oral contraceptives [2], warfarin [2], theophylline [25], and simivastatin [26]; garlic supplement decreased the plasma AUC and C_{max} of a protease inhibitor saquinavir [27]; and piperine increased C_{max} and AUC of phenytoin [28], propranolol, and theophylline [29].

Altered Absorption

Absorption refers to the physical passage of drugs or herbs from the outside to the inside of the body. Most absorption occurs in the intestines where drugs pass through the intestine wall to enter the blood. Several mechanisms may interfere with the intestinal absorption of drugs. Firstly, changes in drug absorption may be due to the alternation of pH of stomach by other drugs. Drugs may neutralize, decrease or increase the secretion of gastric acid. When the secretion of gastric acid is decreased by drugs, acidic drug may not be broken down properly and poor absorption in gut happens. For example, rise in pH due to H_2-blockers and antacids can reduce the absorption of ketoconazole [30], and the absorption of enoxacin is also possibly reduced by rises in pH due to ranitidine [31]. Secondly, drug may bind to other drugs to form an insoluble complex by chelating, absorption or ion exchange, in which the absorption of both substances may decrease because the size of the insoluble complex is too large to pass through the intestinal wall. For example the tetracycline antibiotics can chelate with a number of di- and trivalent metallic ions such as calcium, aluminum, bismuth and iron to form complexes which not only are poorly absorbed but reduce antibacterial effects [32]. Thirdly, drugs that affect gastrointestinal motility may affect the absorption of anther drug. Gastrointestinal motility is the rate at which the intestine contracts to push the content from stomach to the rectum. When one drug causes a decrease in gastrointestinal motility, the interacted drug may stay in the intestine for a longer period to enable an increase in absorption. When one drug speeds up the intestinal transit, the interacted drug may have less time to be absorbed. For example, aloe can cause diarrhea which reduced the absorption of warfarin [33]. Reduced levodopa absorption has been observed with homatropine [34].

Moreover, drugs may be excreted into the gastrointestinal lumen by P-glycoprotein, a product of the multidrug resistant gene that lowers intracellular drug concentrations by acting as an energy-dependent drug efflux pump [35-37]. P-glycoprotein is mainly distributed in large and small intestine, kidney and liver. Being a substrate of P-glycoprotein, the drug may be secreted back to gastrointestinal lumen and less adsorption would be resulted. Drugs or herbal products that inhibit or induce P-glycoprotein may increase or decrease plasma concentration of P-glycoprotein substrate. Thus, P-glycoprotein may be involved in many drug interactions occurring in the gastrointestinal tract, liver, and kidney. For example, the induction of intestinal P-glycoprotein by rifampicin causes the secretion of digoxin back into gastrointestinal tract [38-40].

Altered Distribution

Distribution refers to the process in which drugs are transported to different parts of the body. Interactions occur during the distribution phase if the drug is highly protein-bound, which may be displaced from its binding sites by the interacting drug.

Drug displacement from its protein-bound forms causes an increase in serum drug levels; therefore an increase in the therapeutic effect will be resulted. Herbs containing pain-reducing salicylates, such as meadowsweet and black willow, may displace highly protein-bound drugs such as warfarin, thus increasing the adverse effects of warfarin [41].

Altered Metabolism

The role of metabolism is to change these active lipid-soluble compounds to inactive water-soluble substances that can be efficiently excreted. Hepatic microsomal enzymes oxidize, demethylate, and hydrolyze (phase I) to increase the water solubility as well as add on a functional group for phase II conjugation. Then large water-soluble molecules such as glucuronic acid and sulfate are attached to the drug (phase II) to form the usually inactive water-soluble metabolit.

Cytochrome P450 monooxygenases are responsible for the oxidation of many drugs, such as warfarin, phenytoin, tolbutamide, and cyclosporine. Among the various CYP isozymes, CYP1A2, 2C9, 2C19, 2D6 and 3A4 are major isoforms responsible for more than 90% metabolism of drugs in man [13, 14]. The rate at which the liver metabolize the drugs determines the duration of action of these drugs [42]. An enzyme inducer is a drug that causes an increase in the activity of CYP isoenzymes by increasing their synthesis and metabolic capacity.

Drug-drug interactions occur when an enzyme inducer stimulates enzyme activity to increase metabolism of drug such that a higher dose of drug is required. For example, St. John's wort can induce hepatic CYP isozymes to increase the metabolism of drugs such as digoxin, warfarin, theophylline, and proteases inhibitors [43-45]. Apart from enzyme induction, drug-drug interactions due to the inhibition of CYP isozymes by a drug cause an increase in serum concentration of another drug. If a drug has a narrow therapeutic index, toxicity of the drug will occur.

In addition, metabolism of one drug may be inhibited by another one because both are competing for the same binding sites of the P450 enzymes. CYP isozymes may be saturated when more than two drugs with the same metabolic pathway are administered. This may result in a decrease in the rate of metabolism of one or both drugs. At the same time, certain drugs may bind to CYP isozymes to inhibit enzyme function. For example, cimetidine binds to certain isozymes of CYP [46, 47]. Of the identified CYP genes, families CYP1, CYP2, and CYP3 appear to be involved primarily with drug metabolism [48]. A number of drugs have been identified as substrates, inhibitors, and inducers of metabolism by the CYP enzymes (Table 1-3).

Altered Elimination

The kidney is responsible for eliminating drugs from the body. When blood passes through the glomeruli of kidneys, drugs with low molecules weight can filter through the pores of glomerular membrane. If drugs are small and water soluble or large and lipid soluble, they may be reabsorbed into the body by passive diffusion or active transport. Factors influence the excretion including pH effect and altered active transport in the tubules [34].

Table 1. Substrates of CYP enzymes

CYP	Substrates					
1A2						
	amitriptyline	caffeine	clomipramine	clozapine	cyclobenzaprine	estradiol
	fluvoxamine	haloperidol	imipramine N-DeMe	phenacetin	mexiletine	ropivacaine
	naproxen	acetaminophen	propranolol	riluzole	tacrine	theophylline
	verapamil	(R)warfarin	zileuton	zolmitriptan		
2B6	bupropion	cyclophosphamide	efavirenz	ifosfamide	methadone	
2C19	lansoprazole	omeprazole	pantoprazole	E-3810	phenytoin(O)	
	S-mephenytoin	phenobarbitone	amitriptyline	carisoprodol	citalopram	clomipramine
	cyclophospham ide	hexobarbital	imipramine N-DeME	indomethacin	R-mephobarbital	moclobemide
	nelfinavir	nilutamide	primidone	progesterone	proguanil	propranolol
	teniposide	R-warfarin				
2C9	diclofenac	ibuprofen	meloxicam	S-naproxen	piroxicam	suprofen
	tolbutamide	glipizide	losartan	irbesartan	amitriptyline	celecoxib
	fluoxetine	fluvastatin glyburide	phenytoin	rosiglitazone	tamoxifen	torsemide
	S-warfarin					
2D6	carvedilol	S-metoprolol	propafenone	timolol	amitriptyline	clomipramine
	desipramine	imipramine	paroxetine	haloperidol	perphenazine	risperidone

Table 1. (Continued)

CYP	Substrates					
	thioridazine	alprenolol	amphetamine	bufuralol	chlorpheniramine	chlorpromazine
	debrisoquine	dexfenfluramine	dextromethorphan	encainide	flecainide	fluoxetine
	fluvoxamine	lidocaine	metoclopramide	methoxya-mphetamine	mexiletine	nortriptyline
	minaprine	ondansetron	perhexiline	phenformin	propranolol	quanoxan
	sparteine	tamoxifen	tramadol	venlafaxine		
2E1	enflurane	halothane	isoflurane	methoxyflurane	sevoflurane	acetaminophen
	aniline	benzene	chlorzoxazone	ethanol	N,N-dimethyl formamide	theophylline
3A4,5,7	clarithromycin	erythromycin	alprazolam	quinidine	zolpidem	diazepam
	midazolam	triazolam	cyclosporine	tacrolimus (FK506)	indinavir	nelfinavir
	ritonavir	saquinavir	cisapride	astemizole	chlorpheniramine	terfenidine
	amlodipine	diltiazem	felodipine	lercanidipine	nifedipine	nisoldipine

CYP	Substrates					
	nitrendipine	verapamil	atorvastatin	cerivastatin	lovastatin	simvastatin
	estradiol	hydrocortisone	progesterone	testosterone	alfentanyl	buspirone
	cafergot	cocaine	dapsone	codeine- N-demethylation	dextrome-thorphan	eplerenone
	fentanyl finasteride	gleevec	haloperidol	irinotecan	LAAM	lidocaine
	methadone	odanestron	pimozide	propranolol	quinine	salmeterol
	sildenafil	sirolimus	tamoxifen	taxol	terfenadine	trazodone
	vincristine	zaleplon				

Table 2. Inhibitors of CYP enzymes

CYP	Inhibitors					
1A2	amiodarone	cimetidine	fluoroquinolones	fluvoxamine	furafylline	interferon
	methoxsalen	mibefradil	ticlopidine			
2B6	thiotepa	ticlopidine				
2C19	cimetidine	felbamate	fluoxetine	fluvoxamine	indomethacin	ketoconazole
	lansoprazole	modafinil omeprazole	paroxetine	probenicid	ticlopidine	topiramate
2C9	amiodarone	fluconazole	fluvastatin	fluvoxamine	isoniazid	lovastatin
	paroxetine	phenylbutazone	probenicid	sertraline	sulfamethoxazole	sulfaphenazole
	teniposide	trimethoprim	zafirlukast			
2D6	amiodarone	buproprion	celecoxib	chlorpromazine	chlorpheniramine	cimetidine
	clomipramine	cocaine	doxorubicin	fluoxetine	halofantrine	red-haloperidol
	levomepromazine	metoclopramide	methadone	mibefradil	moclobemide	paroxetine
	quinidine	ranitidine	ritonavir	sertraline	terbinafine	
2E1	dithiocarbamate	disulfiram				
3A4,5,7	delaviridine	indinavir	nelfinavir	ritonavir	saquinavir	amiodarone
	azithromycin	cimetidine	ciprofloxacin	clarithromycin	diethyl- dithiocarbamate	diltiazem
	erythromycin	fluconazole	fluvoxamine	gestodene	grapefruit juice	itraconazole
	ketoconazole	mifepristone	nefazodone	norfloxacin	norfluoxetine	mibefradil
	verapamil					

Table 3. Inducers of CYP enzymes

CYP	Inducers					
1A2	broccoli	brussel sprouts	char-grilled meat	insulin	methyl cholanthrene	modafinil
	nafcillin	beta- naphthoflavone	omeprazole	tobacco		
2B6	phenobarbital	rifampin				
2C19	carbamazepine	norethindrone	pentobarbital	prednisone	rifampin	
2C9	rifampin	secobarbital				
2D6	dexamethasone	rifampin				
2E1	ethanol	isoniazid				
3A4,5,7	efavirenz	nevirapine	barbiturates	carbamazepine	glucocorticoids	modafinil
	phenobarbital	phenytoin	rifampin	St. John's wort	troglitazone	pioglitazone
	rifabutin					

A change in urine pH affects the excretion and the degree of ionization of some drugs. Excretion of weakly acidic drugs is increased in alkaline urine and decreased in acidic urine because weakly acidic drugs will be mainly ionized in highly alkaline urine and reabsorption will be prevented. For example changes in the excretion of quinidine and salicylate due to alterations in urinary pH were caused by antacids [49, 50]. However, most herbs do not change urinary pH to any large extent [18]. Drugs that have the same active transport mechanism may compete for excretion in renal tubules. This competition may reduce renal clearance, prolong the action, and increase accumulation and toxicity of other drugs. For example, probenecid acts in this way to reduce the excretion of penicillins and increase blood concentrations of the latter [51]. In a similar way, salicylates and some NSAIDs can increase methotrexate concentrations possibly leading to serious toxicity [52-56].

Pharmacodynamic Interactions

Pharmacodynamic interactions are effects of one drug that are affected by the presence of another drug at the site of action. This change in drug action does not alter plasma concentration. There is a fluctuation in bioavailability of ingested substances as a result of additive, synergistic or antagonistic interactions between herb and drug molecules [18]. An additive effect occurs when two drugs of similar properties show additive or synergistic increase in clinical effects when given together. For example, alcohol can depress the central nervous system. If taken in moderate amounts with normal therapeutic doses of drugs (e.g. sedatives, hypnotisms), alcohol may cause excessive drowsiness [57, 58]. Herbs such as ginkgo, ginger and garlic with antiplatelet activity may increase the risk of bleeding in patients taking warfarin [59]. An antagonistic effect occurs when two drugs of similar properties acting on the same receptor show lessened or no clinical effect when given together. For example the oral anticoagulants can prolong the blood clotting time by competitively inhibiting the effects of dietary vitamin K. If the intake of vitamin K is increased, the effects of the oral anticoagulant are opposed and the prothrombin time can return to normal, therefore cancelling out the therapeutic benefits of anticoagulant treatment [60-64].

Altering receptor sensitivity may also affect the action of drug. One drug may interact with the active site to modify the intensity of the response through noncompetitive processes. For example, instead of affecting the ADME (absorption, distribution, metabolism and excretion) of the anticoagulant warfarin, d-thyroxine increases the affinity of warfarin receptor site for warfarin [65]. The increase in receptor affinity for warfarin will increase anticoagulant effect in patients [66].

PROBLEMS IN MONITORING DRUG/HERB INTERACTIONS

Recent research has estimated that about 50% of adult Americans use at least one prescription medication and 7% adult Americans take 5 or more prescription drugs [67]. Among the prescription drug users, 16% also take herbal supplements, but the prevalence of clinically significant interactions between herbals and medications is unknown [68]. In China, more than 90% of the population use Chinese herbal medication and some may be taking western drugs for treatment at the same time (China Chinese Medicine's Market Report 2010). However, there is little information on how most herbal products are metabolized or how they affect the liver's ability to metabolize other drugs because manufactures are not required to submit documentation of herbal product testing, their side effects and interactions with other drugs. For example, potential interactions between warfarin and herbal products cannot be predicted with certainty because the pharmacokinetic and pharmacodynamic properties of herbal products are not fully understood [3].

As a complex mixture of substances extracted from plant sources, herbal medicines contain mixtures of pharmacologically active constituents. Generally, a complete characterization of all the chemical constituents in one herb is unknown. Besides, composition of an herbal product may vary with the part of the plant processed (stems, leaves, roots), season and growing condition. Combination products composed of multiple herbal products complicate matters further. Not only does the complex nature of herbal products complicate the determination of herb-drug interactions, but the manufacturing process contributes to the overall complexity as well. Since herbal products are not regulated by the U.S. Food and Drug Administration (FDA) as previously stated, there are no unitive standards for herbal products. In fact, some products have been found to be misidentified, substituted and adulterated with other herbal products [68].

In most cases, the herb-drug interactions depend on factors related to drugs (dose, dosing regimen, administration route, pharmacokinetic and therapeutic range) and patients (genetic polymorphism, age, gender and pathological conditions) [69]. Patients may not inform their doctors of having taken the herbal products because they have a fear of disapproval or they do not consider herbal products as drugs. Therefore, it is especially difficult to assess the incidence of herb-drug interactions and the reliability of these reports which often involve only one patient or is a theoretical case with no evidence.

METHODS TO ASSESS DRUG INTERACTIONS

Recently there has been an increasing use of *in vitro* and *in vivo* models to explore possible drug interactions, resulting from the identification of more clinically relevant herb-drug metabolism interactions, such as St. John's wort-warfarin interactions [12], Danshen-warfarin interactions [70], and ginkgo-warfarin interactions [6, 7].

In Vitro Models

Drug interactions can be investigated by a number of *in vitro* systems, including the use of sub cellular fractions of liver, precision-cut liver slices, isolated hepatocytes or established liver cell lines, and cDNA-expressed enzymes [71-77]. These systems have both advantages and limitations, and a combination of methods will provide more accurate information on the drug interactions. Liver microsomes are easily manipulated, optimized, and are ideal for studying Phase I and II reactions. However, cofactors are necessary for CYP- or uridine diphosphate glucuronotransferase (UGT)-catalyzed reactions to replace those lost due to the destruction of cell integrity. In addition, because of the latter, no coupled metabolism is present, and Phase II reactions following a Phase I reaction cannot be studied. In contrast, hepatocytes provide cellular integrity with respect to enzyme architecture and allow the study of Phase II reactions following Phase I metabolism.

Human and animal hepatocytes have been considered a particularly useful tool to study drug interactions [43, 78, 79]. Apart from metabolic studies, hepatocytes also provide a valuable tool to assess the effects of drugs and/or herbs on human CYPs at the level of protein, mRNA, and enzyme activity. Selective chemical inhibitors for various CYPs have been showed to be useful tools for studying drug interactions (Table 2) [80-85]. Most of these chemical probes are now commercially available. When multiple CYP isozymes are involved in the metabolism of one drug or a compound from herbs, it is a powerful means to demonstrate the relative importance of each isozyme quantitatively. The effectiveness of competitive or reversible inhibitors is dependent on the concentration of both the inhibitor and the substrate (usually at K_m).

The specificity of isozyme-selective inhibitors of CYPs may also be concentration-dependent [83]. One caution needs to keep in mind that primary hepatocyte function, especially CYP isozyme activity declines over time in culture, which may affect to interpret valuable data, and maintaining the integrity of primary hepatocytes has been challenging, and so far less success has been achieved [86].

In Vivo Models

Although *in vitro* models provide a quick screening method to study the metabolism based drug interactions, *in vivo* studies are essential to confirm the relevance and clinical significance. Animal studies may provide important information on drug interactions, but interspecies variations in the substrate specificity, catalytic features, and amino acid sequences of CYPs may cause difficulty in extrapolating animal data to humans [87-89]. Therefore, it may be difficult to predict accurately the effects of herbal preparations in humans based on animal studies, and human studies are usually required to confirm herb–drug interactions.

Probe substrates/inhibitors (see Table 1 and 2) have been extensively used to explore the effects of drugs on specific CYP enzyme activity *in vivo*, e.g., caffeine for CYP1A2 (90), tolbutamide for CYP2C9 [80], mephenytoin for CYP2C19 [76], dextromethorphan, or debrisoquin for CYP2D6 [91], chlorzoxazone for CYP2E1 (92), and midazolam or erythromycin for CYP3A4 [76, 93, 94]. In addition, a cocktail of probe drugs has been used to explore the activities of multiple CYPs in one test system [95-98].

In Silico Methods

Major *in silico* methods include simple rule-based modeling, structure–activity relationships, three-dimensional quantitative structure–activity relationships (QSAR), and pharmacophores [99]. These are useful tools for understanding the CYP-catalyzed reactions and in predicting possible drug interactions with an underlying metabolic mechanism, and in calculating pharmacokinetic parameters such as clearance [99]. The resulting data based on *in silico* approaches may be of clinical relevance and significance.

With the isolation and identification of active constituents of herbal preparations, there has been an increasing use of *in silico* models to study their pharmacological effects. A pharmacophore model with inhibitory effects on platelet activating factor has been established [100]. For example, structure–activity relationship analysis has indicated that the presence of a furano-o-naphthoquinone in tanshinone analogues isolated from the chloroform extract of Danshen roots is the basic requirement for cytotoxic activity to tumor cells [101].

This study has also indicated that the planar phenanthrene ring of the tanshinones may be essential for interaction with the DNA molecule; whereas the furano-o-quinone moiety may be responsible for the production of reactive free radicals in the close vicinity of the bases to cause DNA damage [101].

Recently we also have reported that the structural difference between dihydrotanshinone and tanshinone I is only the presence of double bond at C-15 position of furan ring, but this resulted in the different inhibition of CYP3A [102, 103]. Dihydrotanshinone as a noncompetitive inhibitor binds to an inhibitor site on the CYP3A which is remote from the active site and brings about a conformational change in the active site of CYP3A. The change in the active site prevents the enzyme from converting the bound substrate to product.

Indeed, furans may behave as suicidal inhibitors of enzymes, but in tanshinones the situation is complicated by the presence of a carbonyl adjacent to the furan moiety [102]. The reactivity of the furan ring of tanshinones with electrophiles may result in the different inhibition of CYP3A [102].

CYTOCHROME P450 ENZYMES IN DRUG INTERACTIONS

A drug-drug interaction usually occurs as CYP has broad substrate specificity and overlapping substrate selectivity [12]. In addition, the activity of CYP can be altered by enzyme induction or inhibition, thus affecting the disposition and pharmacological action of co-administered drugs. For example, some clinical studies have documented that St. John's wort reduced the plasma concentration of drug indinavir (substrate of CYP3A4) because of the induction of hepatic and intestinal CYP3A4 by St. John's wort [23]. Therefore, we need to know the background of CYP enzymes in order to understand the role of CYP in drug interactions.

Nomenclature of Cytochrome P450

Based on amino acid sequence homology, CYP superfamily members are divided into various subfamilies. The drug metabolizing CYP enzymes are confined to subfamilies 1, 2, 3, and 4. These subfamilies are further divided into isoforms. A nomenclature for the cytochrome P450 has been developed to help understand their inter-relationships [104].

CYP---- The capital letters 'CYP' indicate that the isoform is of human origin.

CYP 'x' An Arabic numeral is next used to indicate the isoform family.

CYP 'x' 'X' Subfamilies are designated by another capital letter.

CYP 'x' 'X' 'x' A final Arabic numeral designates the individual gene product in the subfamily.

Designation of family and subfamily status relates to the degree of homology for the amino acid sequence of various P450 isoforms. For example, the designation CYP1A2 refers to a human cytochrome P450 isoform of the first family that is a member of the A subfamily of enzymes and is the second gene product assigned to that subfamily. A total of 270 CYP gene families found in various organisms have been described to date [105]. The 18 gene families that exist in humans encode 74 individual CYP genes. Despite the large number of CYP genes and enzymes, it appears that only the CYP1, CYP2, and CYP3 families of enzymes have a major role in drug metabolism [104, 105].

Catalytic Mechanism of CYP Enzymes

Drug oxidation by the CYP system requires drug (substrate, 'DH'), CYP enzyme, molecular oxygen, NADPH and a flavoprotein (NADPH-P450 reductase). The mechanism involves a complex cycle, but the overall effect of the reaction is quite simple, namely the addition of one atom of oxygen to the drug to form a hydroxyl group (product, 'DOH'), the other atom of oxygen being converted to water [15]. The primary steps in the catalytic cycle for the reactions catalyzed by the CYPs are (Figure 1): 1) substrate binding; 2) one-electron reduction of the ferric (Fe^{3+}) enzyme to the ferrous (Fe^{2+}) enzyme; 3) binding of molecular oxygen to the ferrous (Fe^{2+}) iron; 4) transfer of the second electron to the ferrous-oxy-substrate complex causing the release of water and the formation of an activated oxygen intermediate; 5) the catalytic insertion of the activated oxygen into the substrate to form the oxygenated product; and 6) the release of the oxygenated product resulting in the release of the native ferric (Fe^{3+}) form of the enzyme that can then undergo another catalytic cycle.

The nature of the catalytic cycle of the CYP enzymes presents a number of potential points at which inhibition of substrate metabolism may occur. The basic types of enzyme inhibitors include: 1) competitive; 2) noncompetitive; 3) uncompetitive; 4) mixture inhibition. Drugs may inhibit CYPs by three mechanisms: competitive inhibition, noncompetitive inhibition, and mechanism-based inhibition. Mutual competitive inhibition may occur between a drug and one herbal constituent, as both are often metabolized by the same CYP isoform. For example, diallyl disulfide from garlic is a competitive inhibitor of CYP2E1 [106]. Recently we also reported tanshinones competitively inhibited rat CYP1A2 activity and the K_i values were in the order: dihydrotanshinone (3.64 µM), cryptotanshinone (4.07 µM), tanshinone I (22.6 µM) and tanshinone IIA (23.8 µM) [107]. Further study in human liver *in vitro* showed tanshinone I, tanshinone IIA, cryptotanshinone, and dihydrotanshinone were potent competitive CYP1A2 inhibitors with the K_i values at 2.16, 1.45, 1.88 and 0.53 µM, respectively [103].

Noncompetitive inhibition is caused by the binding of drugs or herbal constituents containing electrophilic groups (e.g., imidazole or hydrazine group) to the haem-portion of CYP. For example, piperine inhibited arylhydrocarbon hydroxylase (CYP1A) and 7-ethoxycourmarin deethylase (CYP2A) by noncompetitive mechanism [108, 109]. Hyperforin present in St. John's wort is also a potent noncompetitive inhibitor of CYP2D6 activity [110]. The mechanism-based inhibition of CYP is due to the formation of a complex between drugs with CYP enzymes. For instance, dihydrotanshinone is a potent noncompetitive inhibition toward CYP3A4 activity in human CYP3A4 isoform with IC_{50} of 5.57 µM and K_i of 2.11 µM [103]. The difference between dihydrotanshinone and tanshinone I is in the presence of double bond at C-15 position of furan ring, but this resulted in the difference in inhibition of CYP3A4 activity [102]. The noncompetitive inhibition exhibited by dihydrotanshinone may, however, be related to its binding to another site on the CYP3A4 isoform which is remote from the active site and brings about a conformational change in the active site of CYP3A4 [102]. The change in the active site prevents the enzyme from converting the bound substrate to product.

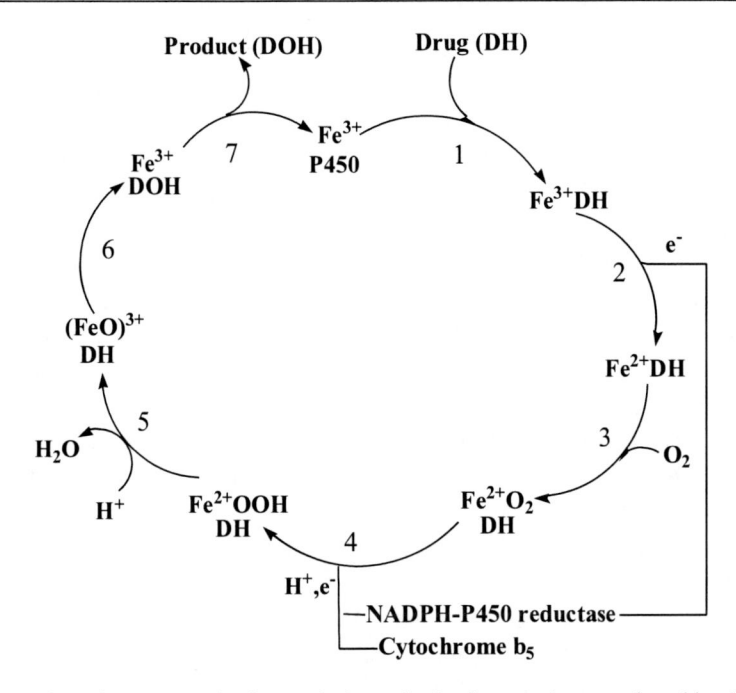

Figure 1. The primary steps in the catalytic cycle for the reactions catalyzed by CYP enzymes.

Probe Drugs of CYP Enzymes

Probe substrates (or probe drugs) refer to compounds that are predominately or exclusively metabolized *in vitro* by an individual CYP enzyme [76]. The metabolism of the candidate probe is generally characterized through the use of preparations containing individually expressed human CYP enzymes or human liver microsomes [111]. Drugs that are selectively metabolized and that can be safely administered to humans may be used as *in vivo* probe drugs to estimate CYP enzyme activity (i.e., phenotyping). For example, the nonsteroidal anti-inflammatory drug flurbiprofen is selectively metabolized *in vitro* by CYP2C9 to 4-hydroxyflurbiprofen [112], providing the necessary justification for exploring the use of flurbiprofen as an *in vivo* probe of CYP2C9 activity [113].

The phenotyping procedure typically involves the administration of the probe drug and the collection of blood and/or urine in order to determine functional activity of the enzyme. Typically, a small dose of the probe drug is administered to avoid or minimize undesirable clinical effects. An index of enzyme activity, also referred to as a phenotypic trait measure, is chosen to reflect the catalytic activity of a single pathway of metabolism. The intrinsic clearance of a probe or of the metabolite(s) produced is the appropriate measure of enzyme activity [114]. For example, the short-acting benzodiazepine midazolam is frequently used as an *in vivo* probe of CYP3A4 activity. Midazolam is predominantly metabolized to 1'-hydroxymidazolam by both CYP3A4 and CYP3A5 (collectively referred to as CYP3A) [115]. After intravenous or oral administration of subtherapeutic and low dose midazolam, multiple blood samples are obtained to measure midazolam clearance, which serves as an index of CYP3A4 activity [116]. Midazolam has been used extensively to characterize other factors (e.g., genetics and drug-drug interactions) that affect CYP3A-mediated metabolism [10, 76, 117-119].

Probe drugs have been used to gain important insights into the clinical relevance of genetic variation (i.e., pharmacogenetics) and to characterize nongenetic factors that influence metabolism, including demographic characteristics (e.g., age, sex, weight, etc.), as well as, drug–drug or drug–herb interactions. A partial list of probe drugs is provided in Table 4. For our evaluation, an ideal probe substrate is the one with a simple metabolic scheme, so that the formation rate of a metabolite specifically reflects the activity of one distinct CYP isoform.

Traditional Chinese Medicine (TCM) has been widely used in the treatment of diseases because its therapeutic efficacy is mild and broad, and the incidence of adverse reactions is relatively low in comparison with synthetic drugs [120].

However, TCM contains many compounds that may influence the activity of CYP isoforms [121]. For example, *Angelica dahurica* extract inhibits various CYP isoforms, such as CYP2C, CYP3A and CYP2D1 [122]. Our previous studies have demonstrated that Danshen aqueous extract affected the metabolism of CYP1A2 substrates through competitive inhibition and altered their clearance in humans and rats [123]. Recently we have reported that Danshen extract in the form of capsules inhibited the liver CYP3A activity in humans and rats *in vitro* [124], prolonged the hypnotic effects of midazolam, reduced the clearance of midazolam, and decreased CYP3A expression in Sprague-Dawley rats [125]. In addition, tanshinones (tanshinone I, tanshinone IIA, dihydrotanshinone, and cryptotanshinone) isolated from Danshen exhibit different modes of inhibition on human CYP1A2, 2C9, 2E1, and 3A4 activity, respectively [103, 126].

**Table 4. Selected *in vitro* and *in vivo* probe substrates
for human CYP enzymes**

CYP	Probe Substrates	
	In Vitro	*In Vivo*
CYP1A2	Phenacetin	Caffeine
CYP2A6	Coumarin	Coumarin
CYP2B6	Bupropion	Bupropion
CYP2C9	Diclofenac	Tolbutamide
	Tolbutamide	S-Warfarin
	S-Warfarin	
CYP2C19	S-Mephenytoin	S-Mephenytoin
	Omeprazole	Omeprazole
CYP2D6	Bufurolol	Dextromethrophan
CYP2E1	Chlorzoxazone	Chlorzoxazone
CYP3A4	Midazolam	Midazolam
	Testosterone	Erythromycin

Modulation of CYP activity may cause pharmacokinetic changes in other drugs, resulting in a decrease in efficacy or increase in side effects [44, 127-129]. As there is an increasing use of TCM in combination with western drugs in the community, it is necessary to study TCM-drug interactions at metabolic basics. *In vitro* and *in vivo* studies have indicated that the active ingredients in herbs or TCM interact with various CYP isoforms extensively by acting as CYP substrate, inducer or inhibitor [44, 67, 68, 130-139].

CYP Enzymes in Drug Interactions

At the level of hepatic drug metabolism, the two mechanisms of drug–drug interactions are induction and inhibition of drug metabolizing enzymes such as CYP enzymes. Inductive drug–drug interactions occur when an inducing drug causes an increase in the metabolism of co-administered drugs with a resultant diminution in their therapeutic effect. The most important CYP enzyme affected by this form of interaction is CYP3A4 because of the potential magnitude of its induction and the vast array of drugs metabolized by CYP enzymes [140]. Only recently the mechanisms of CYP3A4 induction by inducing drugs such as rifampicin, phenytoin, carbamazepine and the herbal remedy St John's Wort have been elucidated. In fact, most CYP3A4-inducing drugs are now recognized to be ligands for PXR, a nuclear receptor that is highly expressed in hepatocytes and intestinal epithelial cells. PXR/RXR heterodimers bind to response elements in the regulatory upstream region of the CYP3A4 gene, resulting in transcriptional activation [141]. Several other drug-metabolizing genes and drug transporters are also induced by ligand-activated PXR, while a lesser number are activated by other receptors, including CAR and the aryl hydrocarbon receptor (AhR) [142, 143].

Inhibition of metabolism by co-administered drugs is also a common cause of drug–drug interactions and occurs by two basic mechanisms. The most common mechanism of inhibition is simple competition between drugs for access to the catalytic pocket of the relevant drug-metabolizing enzyme. Some drugs are highly effective competitors for CYP enzymes, including cimetidine, ketoconazole and indinavir. The second way in which drugs inhibit CYP enzymes often referred to as 'mechanism-based CYP inhibition', involves the formation of a catalytically inactive, covalently bound complex between a metabolite of the substrate drug and the CYP enzymes [12]. Macrolide antibiotics such as erythromycin and clarithromycin as well as tamoxifen, fluoxitine and the anti-HIV agents, ritonavir and delavirdine, are examples of drugs that interact in this way.

In fact, the studies on the CYP3A4 enzyme and drug-drug interactions are becoming an integral part of drug development and research. In recent years, some case reports of serious even fatal reactions due to concomitant administration of certain drugs require careful consideration. Therefore, drug prescribing for patients on multidrug regimens warrants thorough review of the patient's current therapy with respect to drug biotransformation. For CYP3A4-metabolized drugs that require periodic monitoring of serum levels, the interaction of another CYP3A4-metabolized drug can be controlled by dosage adjustments to maintain appropriate levels of the monitored drug. For instance, cyclosporine, tacrolimus, and carbamazepine are all substrates of CYP3A4.

Thus, co-administration of cyclosporine with a CYP3A4 inhibitor should decrease an individual's cyclosporine dosage requirement. Moreover, drinking grapefruit juice may be an inexpensive way to reduce cyclosporine dosages, but the unpredictable nature of the inhibition of cyclosporine metabolism has not vindicated this practice. Furthermore, ketoconazole and diltiazem (CYP3A4 inhibitors) have been used successfully in this respect. Patients unable to obtain therapeutic cyclosporine levels with orally administered cyclosporine due to inadequate absorption can be placed on either of these agents to achieve this goal.

The real problem, however, with prescribing drugs that share the CYP3A4 pathway has been seen with drugs whose levels are not measured. When the serum levels of these drugs reach a toxic state, the toxicity can manifest itself with serious medical consequences. For example, the pro-arrhythmic effects from high serum levels of the nonsedating antihistamines terfenadine and astemizole have severely limited their usefulness and led to the development of newer agents to take their place. Mibefradil, a potent inhibitor of CYP3A4, was withdrawn from the market after numerous reports of serious drug-drug interactions.

Another drug class of note in this category is the 3-hydroxy-3-methylglutaryl-coenzyme A (HMGCA) reductase inhibitors. High serum concentrations of some of these agents are strongly linked to the development of rhabdomyolysis.

Adding a CYP3A4 inhibitor to a drug regimen that includes certain HMGCA reductase inhibitors greatly increases the patient's risk of developing rhabdomyolysis [144]. One advantage of recognizing this drug interaction has been the subsequent studies conducted to identify which agents can be used safely in multidrug combinations.

Research focusing on CYP3A4 inhibitors and HMGCA reductase inhibitors has found that pravastatin and fluvastatin can be co-administered with itraconazole, a potent CYP3A4 inhibitor, without significant changes in maximum serum concentrations [144-146].

CONCLUSION

During the past few years a revolution has taken place in our understanding of drug interactions, mostly as a result of advances in the molecular biology of the CYP enzyme system. This will allow us to make the most appropriate choices in prescribing and avoiding preventable adverse drug reactions or causes of death. Many drug interactions are a result of inhibition or induction of CYP enzymes.

Therefore, clinicians should be familiar with the substrates, inhibitors, and inducers of the common CYP enzymatic pathways responsible for drug metabolism, and be aware of the potential interactions. At the same time, physicians also need to anticipate and manage drug interactions by understanding the unique functions and characteristics of CYP enzymes. In a word, all efforts will enhance the use of rational drug therapy and better drug combinations.

ACKNOWLEDGMENT

This work was partly supported by grants from the National Natural Science Foundation of China (No. 81301908), and the Science and Technology Commission of Shanghai Municipality (No. 13ZR1412600 and 14DZ2270100).

REFERENCES

[1] Ernst E. Second thoughts about safety of St John's wort. *Lancet,* 1999; 354 (9195):2014-6.

[2] Yue Q. Y., Bergquist C., Gerden B. Safety of St John's wort (Hypericum perforatum). *Lancet,* 2000; 355 (9203):576-7.

[3] Chan T. Y. Interaction between warfarin and danshen (Salvia miltiorrhiza). *Ann. Pharmacother.,* 2001; 35 (4):501-4.

[4] Page R. L., 2nd, Lawrence J. D. Potentiation of warfarin by dong quai. *Pharmacotherapy,* 1999; 19 (7):870-6.

[5] Janetzky K., Morreale A. P. Probable interaction between warfarin and ginseng. *Am. J. Health Syst. Pharm.,* 1997; 54 (6):692-3.

[6] Fugh-Berman A. Herb-drug interactions. *Lancet,* 2000; 355 (9198): 134-8.

[7] Fugh-Berman A., Ernst E. Herb-drug interactions: review and assessment of report reliability. *Br. J. Clin. Pharmacol.,* 2001; 52(5): 587-95.

[8] Yu C. M., Chan J. C., Sanderson J. E. Chinese herbs and warfarin potentiation by 'danshen'. *J. Intern. Med.,* 1997; 241 (4): 337-9.

[9] Izzat M. B., Yim A. P., El-Zufari M. H. A taste of Chinese medicine! *Ann. Thorac. Surg.,* 1998; 66 (3): 941-2.

[10] Lamba J. K., Lin Y. S., Schuetz E. G., Thummel K. E. Genetic contribution to variable human CYP3A-mediated metabolism. *Adv. Drug Deliv. Rev.,* 2002; 54(10): 1271-94.

[11] Guengerich F. P. Cytochrome P450: what have we learned and what are the future issues? *Drug Metab. Rev.,* 2004;36(2):159-97.

[12] Zhou S., Yung Chan S., Cher Goh B., et al. Mechanism-based inhibition of cytochrome P450 3A4 by therapeutic drugs. *Clin. Pharmacokinet.,* 2005; 44(3): 279-304.

[13] Chang G. W., Kam P. C. The physiological and pharmacological roles of cytochrome P450 isoenzymes. *Anaesthesia,* 1999; 54(1): 42-50.

[14] Rodrigues A. D. Integrated cytochrome P450 reaction phenotyping: attempting to bridge the gap between cDNA-expressed cytochromes P450 and native human liver microsomes. *Biochem. Pharmacol.,* 1999; 57(5): 465-80.

[15] Rang H. P., Dale M. M., Ritter J. M., Moore P. K. 6th Edition Pharmacology: Churchill Livingstone: An imprint of Elsevier Science; 2007.

[16] Zhou S., Chan E., Li S. C., et al. Predicting pharmacokinetic herb-drug interactions. *Drug Metabol. Drug Interact.,* 2004; 20(3): 143-58.

[17] Zhou S., Huang M., Xu A., Yang H., Duan W., Paxton J. W. Prediction of herb-drug metabolic interactions: a simulation study. *Phytother. Res.,* 2005; 19(6): 464-71.

[18] Williamson E. M. Drug interactions between herbal and prescription medicines. *Drug Saf.,* 2003; 26(15): 1075-92.

[19] Breidenbach T., Hoffmann M. W., Becker T., Schlitt H., Klempnauer J. Drug interaction of St John's wort with cyclosporin. *Lancet,* 2000; 355 (9218): 1912.

[20] Moschella C., Jaber B. L. Interaction between cyclosporine and Hypericum perforatum (St. John's wort) after organ transplantation. *Am. J. Kidney Dis.,* 2001; 38(5): 1105-7.

[21] Johne A., Schmider J., Brockmoller J., et al. Decreased plasma levels of amitriptyline and its metabolites on comedication with an extract from St. John's wort (Hypericum perforatum). *J. Clin. Psychopharmacol.,* 2002; 22(1): 46-54.

[22] Johne A., Brockmoller J., Bauer S., Maurer A., Langheinrich M., Roots I. Pharmacokinetic interaction of digoxin with an herbal extract from St John's wort (Hypericum perforatum). *Clin. Pharmacol. Ther.,* 1999; 66 (4): 338-45.

[23] Piscitelli S. C., Burstein A. H., Chaitt D., Alfaro R. M., Falloon J. Indinavir concentrations and St John's wort. *Lancet,* 2000; 355 (9203): 547-8.

[24] de Maat M. M., Hoetelmans R. M., Math t R. A., et al. Drug interaction between St John's wort and nevirapine. *Aids,* 2001; 15(3): 420-1.

[25] Nebel A., Schneider B. J., Baker R. K., Kroll D. J. Potential metabolic interaction between St. John's wort and theophylline. *Ann. Pharmacother.,* 1999; 33(4): 502.

[26] Sugimoto K., Ohmori M., Tsuruoka S., et al. Different effects of St John's wort on the pharmacokinetics of simvastatin and pravastatin. *Clin. Pharmacol. Ther.*, 2001; 70(6): 518-24.

[27] Piscitelli S. C., Burstein A. H., Welden N., Gallicano K. D., Falloon J. The effect of garlic supplements on the pharmacokinetics of saquinavir. *Clin. Infect. Dis.*, 2002; 34(2): 234-8.

[28] Bano G., Amla V., Raina R. K., Zutshi U., Chopra C. L. The effect of piperine on pharmacokinetics of phenytoin in healthy volunteers. *Planta. Med.*, 1987; 53(6): 568-9.

[29] Bano G., Raina R. K., Zutshi U., Bedi K. L., Johri R. K., Sharma S. C. Effect of piperine on bioavailability and pharmacokinetics of propranolol and theophylline in healthy volunteers. *Eur. J. Clin. Pharmacol.*, 1991; 41(6): 615-7.

[30] Lelawongs P., Barone J. A., Colaizzi J. L., et al. Effect of food and gastric acidity on absorption of orally administered ketoconazole. *Clin. Pharm.*, 1988; 7(3): 228-35.

[31] Grasela T. H., Jr., Schentag J. J., Sedman A. J., et al. Inhibition of enoxacin absorption by antacids or ranitidine. *Antimicrob. Agents Chemother.*, 1989;33(5):615-7.

[32] Dickinson B. D., Altman R. D., Nielsen N. H., Sterling M. L. Drug interactions between oral contraceptives and antibiotics. *Obstet. Gynecol.*, 2001; 98(5 Pt 1): 853-60.

[33] Lutomski D. M., LaFrance R. J., Bower R. H., Fischer J. E. Warfarin absorption after massive small bowel resection. *Am. J. Gastroenterol.*, 1985; 80(2): 99-102.

[34] Stockley H. I. Drug interactions. Fifth Edition ed: Printed in Great Britain by Cambridge University Press, Cambridge; 1999.

[35] Fromm M. F. P-glycoprotein: a defense mechanism limiting oral bioavailability and CNS accumulation of drugs. *Int. J. Clin. Pharmacol. Ther.*, 2000; 38(2): 69-74.

[36] von Moltke L. L., Greenblatt D. J. Drug transporters in psychopharmacology--are they important? *J. Clin. Psychopharmacol.*, 2000; 20(3): 291-4.

[37] Zhang J. T. The multi-structural feature of the multidrug resistance gene product P-glycoprotein: implications for its mechanism of action (hypothesis). *Mol. Membr. Biol.*, 2001; 18(2): 145-52.

[38] Greiner B., Eichelbaum M., Fritz P., et al. The role of intestinal P-glycoprotein in the interaction of digoxin and rifampin. *J. Clin. Invest.*, 1999; 104(2): 147-53.

[39] Fromm M. F., Kauffmann H. M., Fritz P., et al. The effect of rifampin treatment on intestinal expression of human MRP transporters. *Am. J. Pathol.*, 2000; 157(5): 1575-80.

[40] Matheny C. J., Lamb M. W., Brouwer K. R., Pollack G. M. Pharmacokinetic and pharmacodynamic implications of P-glycoprotein modulation. *Pharmacotherapy,* 2001; 21(7): 778-96.

[41] Smith S. A., Kraft S. L., Lewis D. C., Freeman L. C. Plasma pharmacokinetics of warfarin enantiomers in cats. *J. Vet. Pharmacol. Ther.,* 2000; 23(6): 329-37.

[42] Pelkonen O. Human CYPs: in vivo and clinical aspects. *Drug Metab. Rev.,* 2002; 34(1-2): 37-46.

[43] Moore L. B., Goodwin B., Jones S. A., et al. St. John's wort induces hepatic drug metabolism through activation of the pregnane X receptor. *Proc. Natl. Acad. Sci. U S A,* 2000; 97(13): 7500-2.

[44] Zhou S., Gao Y., Jiang W., Huang M., Xu A., Paxton J. W. Interactions of herbs with cytochrome P450. *Drug Metab. Rev.,* 2003; 35(1): 35-98.

[45] Hu Z., Yang X., Ho P. C., et al. Herb-drug interactions: a literature review. *Drugs,* 2005; 65(9): 1239-82.

[46] Levine M., Bellward G. D. Effect of cimetidine on hepatic cytochrome P450: evidence for formation of a metabolite-intermediate complex. *Drug Metab. Dispos.,* 1995; 23(12): 1407-11.

[47] Faux S. P., Combes R. D. Interaction of cimetidine with cytochrome P450 and effect on mixed-function oxidase activities of liver microsomes. *Hum. Exp. Toxicol.,* 1993; 12(2): 147-52.

[48] Schenkman J. B., Griem H. Cytochrome P450: Heidelberg: Springer-Verlag; 1993.

[49] Remon J. P., Van Severen R., Braeckman P. [Interaction between anti-arrhythmics, antacids and antidiarrheals. III. Effect of antacids and antidiarrheals on the in vitro absorption of quinidine salts]. *Pharm. Acta. Helv.,* 1979; 54(1): 19-22.

[50] Shastri R. A. Effect of antacids on salicylate kinetics. *Int. J. Clin. Pharmacol. Ther. Toxicol.,* 1985; 23(9): 480-4.

[51] Gibaldi M., Schwartz M. A. Apparent effect of probenecid on the distribution of penicillins in man. Clin Pharmacol Ther 1968;9(3):345-9.

[52] Mandel MA. The synergistic effect of salicylates on methotrexate toxicity. *Plast. Reconstr. Surg.,* 1976; 57(6): 733-7.

[53] Badr M. Z., Chen T. S. Potentiation of methotrexate-induced gastrointestinal toxicity by non-steroidal anti-inflammatory drugs (NSAIDs) and vincristine. *Toxicology,* 1985; 34(4): 333-40.

[54] Brouwers J. R., de Smet P. A. Pharmacokinetic-pharmacodynamic drug interactions with nonsteroidal anti-inflammatory drugs. *Clin. Pharmacokinet.*, 1994; 27(6): 462-85.

[55] Uwai Y., Saito H., Inui K. Interaction between methotrexate and nonsteroidal anti-inflammatory drugs in organic anion transporter. *Eur. J. Pharmacol.*, 2000; 409(1): 31-6.

[56] Davies N. M., McLachlan A. J., Day R. O., Williams K. M. Clinical pharmacokinetics and pharmacodynamics of celecoxib: a selective cyclo-oxygenase-2 inhibitor. *Clin. Pharmacokinet.*, 2000; 38(3): 225-42.

[57] Mattila M. J., Mattila-Evenden M. E. Effects of alcohol and hypnosedative drugs on digit-symbol substitution: comparison of two different computerized tests. *J. Psychopharmacol.*, 1997;11(4):313-7.

[58] Mattila M. J., Vanakoski J., Kalska H., Seppala T. Effects of alcohol, zolpidem, and some other sedatives and hypnotics on human performance and memory. *Pharmacol. Biochem. Behav.*, 1998; 59(4): 917-23.

[59] Vaes L. P., Chyka P. A. Interactions of warfarin with garlic, ginger, ginkgo, or ginseng: nature of the evidence. *Ann. Pharmacother.*, 2000; 34(12): 1478-82.

[60] Ford S. K., Misita C. P., Shilliday B. B., Malone R. M., Moore C. G., Moll S. Prospective study of supplemental vitamin K therapy in patients on oral anticoagulants with unstable international normalized ratios. *J. Thromb. Thrombolysis*, 2007.

[61] Jie K. S., Gijsbers B. L., Knapen M. H., Hamulyak K., Frank H. L., Vermeer C. Effects of vitamin K and oral anticoagulants on urinary calcium excretion. *Br. J. Haematol.*, 1993; 83(1): 100-4.

[62] Vermeer C., Hamulyak K. Pathophysiology of vitamin K-deficiency and oral anticoagulants. *Thromb. Haemost.*, 1991; 66(1): 153-9.

[63] Gonmori H., Takeuchi T., Tanaka H., Kobayashi N., Maekawa T. Oral anticoagulants and the metabolism of vitamin K-dependent clotting factors. *Nippon Ketsueki Gakkai Zasshi*, 1982; 45(4): 817-29.

[64] Brozovic M. Annotation: Oral anticoagulants, vitamin K and prothrombin complex factors. *Br. J. Haematol.*, 1976;32(1):9-12.

[65] Solomon H. M., Schrogie J. J. Change in receptor site affinity: a proposed explanation for the potentiating effect of D-thyroxine on the anticoagulant response to warfarin. *Clin. Pharmacol. Ther.*, 1967; 8(6): 797-9.

[66] Weintraub M., Breckenridge R. T., Griner P. F. The effects of dextrothyroxine on the kinetics of prothrombin activity: proposed

mechanism of the potentiation of warfarin by D-thyroxine. *J. Lab. Clin. Med.*, 1973; 81(2): 273-9.

[67] Pal D., Mitra A. K. MDR- and CYP3A4-mediated drug-herbal interactions. *Life Sci.,* 2006; 78(18): 2131-45.

[68] Venkataramanan R., Komoroski B., Strom S. In vitro and in vivo assessment of herb drug interactions. *Life Sci.,* 2006; 78(18): 2105-15.

[69] Dresser G. K., Spence J. D., Bailey D. G. Pharmacokinetic-pharmacodynamic consequences and clinical relevance of cytochrome P450 3A4 inhibition. *Clin. Pharmacokinet.*, 2000; 38(1): 41-57.

[70] Chan K., Lo A. C., Yeung J. H., Woo K. S. The effects of Danshen (Salvia miltiorrhiza) on warfarin pharmacodynamics and pharmacokinetics of warfarin enantiomers in rats. *J. Pharm. Pharmacol.*, 1995; 47(5): 402-6.

[71] Eddershaw P. J., Dickins M. Advances in drug metabolism screening. *Pharm. Sci. Technolo. Today*, 1999; 2(1): 13-9.

[72] Li A. P., Gorycki P. D., Hengstler J. G., et al. Present status of the application of cryopreserved hepatocytes in the evaluation of xenobiotics: consensus of an international expert panel. *Chem. Biol. Interact.*, 1999; 121(1): 117-23.

[73] Ekins S., Ring B. J., Grace J., McRobie-Belle D. J., Wrighton S. A. Present and future in vitro approaches for drug metabolism. *J. Pharmacol. Toxicol. Methods,* 2000; 44(1): 313-24.

[74] LeCluyse E. L. Human hepatocyte culture systems for the in vitro evaluation of cytochrome P450 expression and regulation. *Eur. J. Pharm. Sci.,* 2001; 13(4): 343-68.

[75] Rodrigues A. D. Use of in vitro human metabolism studies in drug development. An industrial perspective. *Biochem. Pharmacol.,* 1994; 48(12): 2147-56.

[76] Streetman D. S., Bertino J. S., Jr., Nafziger A. N. Phenotyping of drug-metabolizing enzymes in adults: a review of in-vivo cytochrome P450 phenotyping probes. *Pharmacogenetics*, 2000; 10(3): 187-216.

[77] Venkatakrishnan K., Von Moltke L. L., Greenblatt D. J. Human drug metabolism and the cytochromes P450: application and relevance of in vitro models. *J. Clin. Pharmacol.,* 2001; 41(11): 1149-79.

[78] Goodwin B., Moore L. B., Stoltz C. M., McKee D. D., Kliewer S. A. Regulation of the human CYP2B6 gene by the nuclear pregnane X receptor. *Mol. Pharmacol.*, 2001; 60(3): 427-31.

[79] Wentworth J. M., Agostini M., Love J., Schwabe J. W., Chatterjee V. K. St John's wort, a herbal antidepressant, activates the steroid X receptor. *J. Endocrinol.*, 2000; 166(3): R11-6.

[80] Bourrie M., Meunier V., Berger Y., Fabre G. Cytochrome P450 isoform inhibitors as a tool for the investigation of metabolic reactions catalyzed by human liver microsomes. *J. Pharmacol. Exp. Ther.*, 1996;277(1): 321-32.

[81] Clarke S. E. In vitro assessment of human cytochrome P450. *Xenobiotica*, 1998; 28(12): 1167-202.

[82] Halpert J. R., Guengerich F. P., Bend J. R., Correia M. A. Selective inhibitors of cytochromes P450. *Toxicol. Appl. Pharmacol.*, 1994; 125(2): 163-75.

[83] Newton D. J., Wang R. W., Lu A. Y. Cytochrome P450 inhibitors. Evaluation of specificities in the in vitrometabolism of therapeutic agents by human liver microsomes. *Drug Metab. Dispos.*, 1995; 23(1): 154-8.

[84] Murray M., Reidy G. F. Selectivity in the inhibition of mammalian cytochromes P-450 by chemical agents. *Pharmacol. Rev.*, 1990; 42(2): 85-101.

[85] Pelkonen O., Maenpaa J., Taavitsainen P., Rautio A., Raunio H. Inhibition and induction of human cytochrome P450 (CYP) enzymes. *Xenobiotica*, 1998; 28(12): 1203-53.

[86] Yamamoto N., Wu J., Zhang Y., et al. An optimal culture condition maintains human hepatocyte phenotype after long-term culture. *Hepatol. Res.*, 2006; 35(3): 169-77.

[87] Boobis A. R., Sesardic D., Murray B. P., et al. Species variation in the response of the cytochrome P-450-dependent monooxygenase system to inducers and inhibitors. *Xenobiotica*, 1990; 20(11): 1139-61.

[88] Lewis D. F., Ioannides C., Parke D. V. Cytochromes P450 and species differences in xenobiotic metabolism and activation of carcinogen. *Environ. Health Perspect.*, 1998; 106(10): 633-41.

[89] Lin J. H. Species similarities and differences in pharmacokinetics. *Drug Metab. Dispos.*, 1995;23(10):1008-21.

[90] Carrillo J. A., Christensen M., Ramos S. I., et al. Evaluation of caffeine as an in vivo probe for CYP1A2 using measurements in plasma, saliva, and urine. *Ther. Drug Monit.*, 2000; 22(4): 409-17.

[91] Wieling J., Tamminga W. J., Sakiman E. P., Oosterhuis B., Wemer J., Jonkman J. H. Evaluation of analytical and clinical performance of a

dual-probe phenotyping method for CYP2D6 polymorphism and CYP3A4 activity screening. *Ther. Drug Monit.*, 2000; 22(4): 486-96.

[92] Lucas D., Ferrara R., Gonzalez E., et al. Chlorzoxazone, a selective probe for phenotyping CYP2E1 in humans. *Pharmacogenetics,* 1999; 9(3): 377-88.

[93] Rivory L. P., Slaviero K. A., Hoskins J. M., Clarke S. J. The erythromycin breath test for the prediction of drug clearance. *Clin. Pharmacokinet.*, 2001; 40(3): 151-8.

[94] Brockmoller J., Roots I. Assessment of liver metabolic function. Clinical implications. *Clin. Pharmacokinet.*, 1994; 27(3): 216-48.

[95] Adedoyin A., Frye R. F., Mauro K., Branch R. A. Chloroquine modulation of specific metabolizing enzymes activities: investigation with selective five drug cocktail. *Br. J. Clin. Pharmacol.,* 1998; 46(3): 215-9.

[96] Dierks E. A., Stams K. R., Lim H. K., Cornelius G., Zhang H., Ball S. E. A method for the simultaneous evaluation of the activities of seven major human drug-metabolizing cytochrome P450s using an in vitro cocktail of probe substrates and fast gradient liquid chromatography tandem mass spectrometry. *Drug Metab. Dispos.*, 2001; 29(1): 23-9.

[97] Frye R. F., Matzke G. R., Adedoyin A., Porter J. A., Branch R. A. Validation of the five-drug "Pittsburgh cocktail" approach for assessment of selective regulation of drug-metabolizing enzymes. *Clin. Pharmacol. Ther.,* 1997; 62(4): 365-76.

[98] Zhu B., Ou-Yang D. S., Chen X. P., et al. Assessment of cytochrome P450 activity by a five-drug cocktail approach. *Clin. Pharmacol. Ther.,* 2001; 70(5): 455-61.

[99] Ekins S., Wrighton S. A. Application of in silico approaches to predicting drug--drug interactions. *J. Pharmacol. Toxicol. Methods,* 2001; 45(1): 65-9.

[100] Chen Z. N. [Recent progress in the study of anti-PAF (platelet activating factor) constituents of medicinal herbs]. *Zhongguo Zhong Yao Za Zhi,* 1993; 18(6): 323-6, 61, 80.

[101] Wu W. L., Chang W. L., Chen C. F. Cytotoxic activities of tanshinones against human carcinoma cell lines. *Am. J. Chin. Med.,* 1991;19 (3-4):207-16.

[102] Wang X., Yeung J. H. Inhibitory effect of tanshinones on rat CYP3A2 and CYP2C11 activity and its structure-activity relationship. *Fitoterapia,* 2011; 82(4): 539-45.

[103] Wang X., Cheung C. M., Lee W. Y., Or P. M., Yeung J. H. Major tanshinones of Danshen (Salvia miltiorrhiza) exhibit different modes of inhibition on human CYP1A2, CYP2C9, CYP2E1 and CYP3A4 activities in vitro. *Phytomedicine*, 2010; 17(11): 868-75.

[104] Nelson D. R., Koymans L., Kamataki T., et al. P450 superfamily: update on new sequences, gene mapping, accession numbers and nomenclature. *Pharmacogenetics,* 1996; 6(1): 1-42.

[105] Nebert D. W., Russell D. W. Clinical importance of the cytochromes P450. *Lancet*, 2002; 360(9340): 1155-62.

[106] Teyssier C., Guenot L., Suschetet M., Siess M. H. Metabolism of diallyl disulfide by human liver microsomal cytochromes P-450 and flavin-containing monooxygenases. *Drug Metab. Dispos.*, 1999; 27(7): 835-41.

[107] Wang X., Lee W. Y., Or P. M., Yeung J. H. Effects of major tanshinones isolated from Danshen (Salvia miltiorrhiza) on rat CYP1A2 expression and metabolism of model CYP1A2 probe substrates. *Phytomedicine,* 2009; 16(8): 712-25.

[108] Dalvi R. R., Dalvi P. S. Comparison of the effects of piperine administered intragastrically and intraperitoneally on the liver and liver mixed-function oxidases in rats. *Drug Metabol. Drug Interact.,* 1991; 9(1): 23-30.

[109] Dalvi R. R., Dalvi P. S. Differences in the effects of piperine and piperonyl butoxide on hepatic drug-metabolizing enzyme system in rats. *Drug Chem. Toxicol.,* 1991; 14(1-2): 219-29.

[110] Obach R. S. Inhibition of human cytochrome P450 enzymes by constituents of St. John's Wort, an herbal preparation used in the treatment of depression. *J. Pharmacol. Exp. Ther.*, 2000; 294(1): 88-95.

[111] Bjornsson T. D., Callaghan J. T., Einolf H. J., et al. The conduct of in vitro and in vivo drug-drug interaction studies: a Pharmaceutical Research and Manufacturers of America (PhRMA) perspective. *Drug Metab. Dispos.*, 2003; 31(7): 815-32.

[112] Tracy T. S., Marra C., Wrighton S. A., Gonzalez F. J., Korzekwa K. R. Studies of flurbiprofen 4'-hydroxylation. Additional evidence suggesting the sole involvement of cytochrome P450 2C9. *Biochem. Pharmacol.*, 1996; 52(8): 1305-9.

[113] Lee C. R., Pieper J. A., Frye R. F., Hinderliter A. L., Blaisdell J. A., Goldstein J. A. Tolbutamide, flurbiprofen, and losartan as probes of CYP2C9 activity in humans. *J. Clin. Pharmacol.*, 2003; 43(1): 84-91.

[114] Wilkinson G. R. Clearance approaches in pharmacology. *Pharmacol. Rev.,* 1987; 39(1): 1-47.

[115] Thummel K. E., Wilkinson G. R. In vitro and in vivo drug interactions involving human CYP3A. *Annu. Rev. Pharmacol. Toxicol.*, 1998; 38: 389-430.

[116] Lee J. I., Chaves-Gnecco D., Amico J. A., Kroboth P. D., Wilson J. W., Frye R. F. Application of semisimultaneous midazolam administration for hepatic and intestinal cytochrome P450 3A phenotyping. *Clin. Pharmacol. Ther.*, 2002; 72(6): 718-28.

[117] Yamano K., Yamamoto K., Kotaki H., Sawada Y., Iga T. Quantitative prediction of metabolic inhibition of midazolam by itraconazole and ketoconazole in rats: implication of concentrative uptake of inhibitors into liver. *Drug Metab. Dispos.*, 1999; 27(3): 395-402.

[118] Watanabe M., Tateishi T., Asoh M., et al. Effects of glucocorticoids on pharmacokinetics and pharmacodynamics of midazolam in rats. *Life Sci.*, 1998; 63(19): 1685-92.

[119] Thummel K. E., Shen D. D., Podoll T. D., et al. Use of midazolam as a human cytochrome P450 3A probe: II. Characterization of inter- and intraindividual hepatic CYP3A variability after liver transplantation. *J. Pharmacol. Exp. Ther.*, 1994; 271(1): 557-66.

[120] Lee K. H. Research and future trends in the pharmaceutical development of medicinal herbs from Chinese medicine. *Public Health Nutr.*, 2000; 3(4A): 515-22.

[121] Hodek P., Trefil P., Stiborova M. Flavonoids-potent and versatile biologically active compounds interacting with cytochromes P450. *Chem. Biol. Interact.*, 2002; 139(1): 1-21.

[122] Ishihara K., Kushida H., Yuzurihara M., et al. Interaction of drugs and Chinese herbs: pharmacokinetic changes of tolbutamide and diazepam caused by extract of Angelica dahurica. *J. Pharm. Pharmacol.*, 2000; 52(8): 1023-9.

[123] Wang X., Yeung J. H. Effects of the aqueous extract from Salvia miltiorrhiza Bunge on caffeine pharmacokinetics and liver microsomal CYP1A2 activity in humans and rats. *J. Pharm. Pharmacol.*, 2010; 62(8): 1077-83.

[124] Wang X., Yeung J. H. Effects of Salvia miltiorrhiza Extract on the Liver CYP3A Activity in Humans and Rats. *Phytother. Res.*, 2011; 25:1653-9.

[125] Wang X., Lee W. Y., Zhou X., Or P. M., Yeung J. H. A pharmacodynamic-pharmacokinetic (PD-PK) study on the effects of Danshen (Salvia miltiorrhiza) on midazolam, a model CYP3A probe substrate, in the rat. *Phytomedicine,* 2010; 17(11): 876-83.

[126] Qiu F., Zhang R., Sun J., et al. Inhibitory effects of seven components of danshen extract on catalytic activity of cytochrome P450 enzyme in human liver microsomes. *Drug Metab. Dispos.*, 2008; 36(7): 1308-14.

[127] Homma M., Oka K., Ikeshima K., et al. Different effects of traditional Chinese medicines containing similar herbal constituents on prednisolone pharmacokinetics. *J. Pharm. Pharmacol.*, 1995; 47(8): 687-92.

[128] Wrighton S. A., Stevens J. C. The human hepatic cytochromes P450 involved in drug metabolism. *Crit. Rev. Toxicol.*, 1992; 22(1): 1-21.

[129] Scott G. N., Elmer G. W. Update on natural product--drug interactions. *Am. J. Health Syst. Pharm.*, 2002; 59(4): 339-47.

[130] Hellum B. H., Hu Z., Nilsen O. G. The induction of CYP1A2, CYP2D6 and CYP3A4 by six trade herbal products in cultured primary human hepatocytes. *Basic Clin. Pharmacol. Toxicol.*, 2007; 100(1): 23-30.

[131] Tang J. C., Zhang J. N., Wu Y. T., Li Z. X. Effect of the water extract and ethanol extract from traditional Chinese medicines Angelica sinensis (Oliv.) Diels, Ligusticum chuanxiong Hort. and Rheum palmatum L. on rat liver cytochrome P450 activity. *Phytother. Res.*, 2006; 20(12): 1046-51.

[132] Kuo Y. H., Lin Y. L., Don M. J., Chen R. M., Ueng Y. F. Induction of cytochrome P450-dependent monooxygenase by extracts of the medicinal herb Salvia miltiorrhiza. *J. Pharm. Pharmacol.*, 2006; 58(4): 521-7.

[133] Gurley B. J., Gardner S. F., Hubbard M. A., et al. In vivo effects of goldenseal, kava kava, black cohosh, and valerian on human cytochrome P450 1A2, 2D6, 2E1, and 3A4/5 phenotypes. *Clin. Pharmacol. Ther.*, 2005; 77(5): 415-26.

[134] Zhou S., Koh H. L., Gao Y., Gong Z. Y., Lee E. J. Herbal bioactivation: the good, the bad and the ugly. *Life Sci.*, 2004; 74(8): 935-68.

[135] Delgoda R., Westlake A. C. Herbal interactions involving cytochrome p450 enzymes: a mini review. *Toxicol. Rev.*, 2004; 23(4): 239-49.

[136] Sugiyama T., Kubota Y., Shinozuka K., Yamada S., Yamada K., Umegaki K. Induction and recovery of hepatic drug metabolizing enzymes in rats treated with Ginkgo biloba extract. *Food Chem. Toxicol.*, 2004;42(6):953-7.

[137] Ueng Y. F., Kuo Y. H., Peng H. C., et al. Diterpene quinone tanshinone IIA selectively inhibits mouse and human cytochrome p4501A2. *Xenobiotica*, 2003; 33(6): 603-13.

[138] Guerra M. C., Speroni E., Broccoli M., et al. Comparison between chinese medical herb Pueraria lobata crude extract and its main isoflavone puerarin antioxidant properties and effects on rat liver CYP-catalysed drug metabolism. *Life Sci.*, 2000; 67(24): 2997-3006.

[139] Markowitz J. S., DeVane C. L., Boulton D. W., Carson S. W., Nahas Z., Risch S. C. Effect of St. John's wort (Hypericum perforatum) on cytochrome P-450 2D6 and 3A4 activity in healthy volunteers. *Life Sci.*, 2000; 66(9): PL133-9.

[140] Liddle C., Robertson G. R. Predicting inductive drug-drug interactions. *Pharmacogenomics*, 2003; 4(2): 141-52.

[141] Goodwin B., Hodgson E., Liddle C. The orphan human pregnane X receptor mediates the transcriptional activation of CYP3A4 by rifampicin through a distal enhancer module. *Mol. Pharmacol.*, 1999; 56(6): 1329-39.

[142] Maglich J. M., Stoltz C. M., Goodwin B., Hawkins-Brown D., Moore J. T., Kliewer S. A. Nuclear pregnane x receptor and constitutive androstane receptor regulate overlapping but distinct sets of genes involved in xenobiotic detoxification. *Mol. Pharmacol.*, 2002; 62(3): 638-46.

[143] Whitlock J. P., Jr. Induction of cytochrome P4501A1. *Annu. Rev. Pharmacol. Toxicol.*, 1999; 39: 103-25.

[144] Ogu C. C., Maxa J. L. Drug interactions due to cytochrome P450. *Proc. (Bayl. Univ. Med. Cent.)*, 2000; 13(4): 421-3.

[145] Neuvonen P. J., Kantola T., Kivisto K. T. Simvastatin but not pravastatin is very susceptible to interaction with the CYP3A4 inhibitor itraconazole. *Clin. Pharmacol. Ther.*, 1998; 63(3): 332-41.

[146] Kivisto K. T., Kantola T., Neuvonen P. J. Different effects of itraconazole on the pharmacokinetics of fluvastatin and lovastatin. *Br. J. Clin. Pharmacol.*, 1998; 46(1): 49-53.

In: Cytochrome P450 Enzymes
Editor: Jian Wu

ISBN: 978-1-61942-209-4
© 2014 Nova Science Publishers, Inc.

Chapter 4

CYTOCHROME P450 ENZYMES IN ATHEROSCLEROSIS

Iryna Voloshyna and Allison B. Reiss
Winthrop University Hospital, Mineola, NY, US

ABSTRACT

Cytochrome P450 (P450 or CYP) enzymes are heme-containing enzymes expressed in humans and most living organisms. Cytochrome P450 enzymes were thought to function primarily in drug metabolism until the 1970s when P450 induction was associated with elevation of plasma high-density lipoprotein-cholesterol and apolipoprotein (apo) AI, indicating a link to reduced risk of atherosclerosis. More than 50 human P450 genes have been identified to date.

P450 enzymes provide a front line in maintaining cellular cholesterol homeostasis by augmenting synthesis of hydroxycholesterols (oxysterols) and activating cholesterol-eliminating mechanisms in response to cholesterol accumulation. Major enzymes involved in generation of oxysterols in arterial endothelium and macrophages are CYP7A1, CYP27A1, CYP46A1 and CYP3A4.

P450 cholesterol 27-hydroxylase (CYP27A1) is a key enzyme involved in lipid metabolism and reverse cholesterol transport. The products of CYP27A1 are 27-hydroxycholesterol and 3β-hydroxy-5-cholestenoic acid, polar cholesterol metabolites that are transported to the liver for excretion more readily than cholesterol. The oxysterol 27-hydroxycholesterol behaves like a statin, potently inhibiting HMG CoA reductase while also suppressing smooth muscle cell proliferation and diminishing macrophage foam cell formation. Elevation of cholesterol induces P450 enzymes leading to enhanced production of oxysterols that, in turn, activate genes coding for proteins that transport cholesterol out of cells to the liver for excretion. 27-hydroxycholesterol is an endogenous ligand for the liver X receptor (LXR) α, a nuclear receptor that regulates cholesterol balance. Ligation of nuclear receptors, LXRα and peroxisome-proliferator-activated receptor (PPAR) γ, promote cholesterol efflux from cells of the arterial wall to extracellular acceptors by elevating ATP binding cassette transporters (ABC) A1 and ABCG1 expression. On the other hand, PPARγ agonists may boost the level and activity of 27-hydroxylase, providing one possible mechanism for PPARγ atheroprotection. In addition, overexpression of 27-hydroxylase is associated with increased efflux of cholesterol to apoA-1, independent of ABCA1, by yet undefined mechanisms. Alterations in 27-hydroxylase expression promote cholesterol elimination in the cells of the arterial wall both directly (by catabolism of cholesterol) and indirectly (by modulation of ABCA1 and apoE expression).

This chapter highlights the role of cytochrome P450 enzymes and their metabolic products in atherosclerotic cardiovascular disease, with particular emphasis on their role in cholesterol trafficking. P450 enzyme involvement in bile acid synthesis will also be addressed. Implications for prediction, prevention and treatment of cardiovascular disease will be discussed.

Keywords: Atherosclerosis, Cytochrome P450, High-density lipoproteins, Liver X receptor, Oxysterols, Peroxisome-proliferator-activated receptor, Pregnane X receptor, Reverse Cholesterol Transport

INTRODUCTION

Cytochrome P450 (P450 or CYP) enzymes are heme-containing enzymes expressed in all organisms from bacteria to mammals. Until the 1970s, the function of cytochrome P450 enzymes was connected primarily to drug metabolism. At the present time, CYP enzymes are linked to the biosynthesis of cholesterol, steroids, eicosanoids, bile acids, fatty acids, biogenic amines and vitamin D3 [1]. Alterations in P450 activity can lead to disease states: congenital adrenal hyperplasia, cerebral cholesterinosis; cerebrotendinous xanthomatosis and hypercholesterolemia [2].

Induction of P450 enzyme activity is associateed with elevated plasma high-density lipoprotein-cholesterol and apolipoprotein AI, linking these enzymes to reduced risk of atherosclerosis [3-5]. P450 enzymes provide a front line in maintaining cellular cholesterol homeostasis by augmenting synthesis of hydroxycholesterols (oxysterols) and activating cholesterol-eliminating mechanisms in response to cholesterol accumulation (Figure 1). Major enzymes involved in generation of oxysterols in arterial endothelium and macrophages are CYP7A1, CYP27A1, CYP46A1 and CYP3A4. Evidence gathered over the last decade supports the statement that P450 cholesterol 27-hydroxylase (CYP27A1) is a key enzyme involved in lipid metabolism and reverse cholesterol transport [6-12]. CYP27A1 activity is an important defense mechanism against cholesterol accumulation in macrophages and smooth muscle cells. The importance of CYP27A1 in the control of cholesterol homeostasis and protection against atherosclerosis is supported by the fact that functional deficiency of the enzyme in humans (cerebrotendinous xanthomatosis) is associated with an increased risk of developing premature atherosclerosis [13, 14].

Products of CYP27A1 are 27-hydroxycholesterol and 3β-hydroxy-5-cholestenoic acid, polar cholesterol metabolites that are transported to the liver for excretion more readily than cholesterol [10]. The oxysterol 27-hydroxycholesterol behaves like a statin, potently inhibiting HMG CoA reductase while also suppressing smooth muscle cell proliferation and diminishing macrophage foam cell formation [6, 8, 10]. Endogenously generated oxysterols, such as 20(S)-hydroxycholesterol, 22(R)-hydroxycholesterol, and 24(S), 25-epoxycholesterol, 24- 25- and 27-hydroxycholesterol represent a family of low affinity ligands for the liver X receptor (LXR) α [15-18]. LXRα activates genes encoding transporter proteins that shuttle intracellular cholesterol out of the cell for elimination.

The balanced flow of cholesterol into and out of the macrophage is necessary to avoid lipid overload, and ultimately, atheroma development [19]. Processes that contribute to the elimination of cholesterol from macrophages involve both outflow of cholesterol to extracellular acceptors and intracellular catabolism to polar sterols, exported more readily than cholesterol. Cholesterol efflux from cells of the arterial wall to extracellular acceptors involves the ATP binding cassette transporters (ABC) A1 and ABCG1 [20] (Figure 1). ABCA1 promotes the transfer of cholesterol and phospholipids to lipid-poor apolipoprotein (apo) A-1 [20, 21] contributing to the formation of high-density lipoproteins (HDL) in the liver. ABCG1 is critically involved in regulation of lipid-trafficking mechanisms in macrophages and participates in net efflux of cellular free cholesterol and phospholipid to lipid-free HDL [17]. In addition, macrophages secrete apolipoprotein E (apoE), another cholesterol acceptor that contributes to cholesterol removal [19, 22].

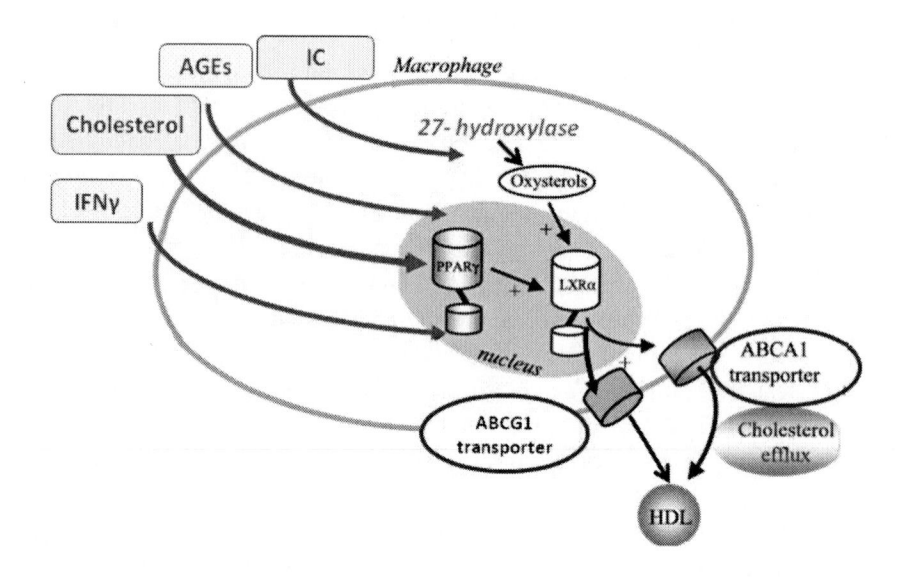

Figure 1. Scheme of alterations in 27-hydroxylase expression in human macrophages. Different agents such as cholesterol loading, immune complexes (IC), pro-inflamatory cytokine IFN-γ, advanced glycation end products (AGEs) affect expression and activity of 27-hydroxylase. 27-Hydroxylase signaling involves PPARγ, LXR/RXRα and impaired signaling leads to downregulation of ABC transporters A1 and G1 and compromises outflow of cholesterol. This creates an imbalance in lipid processing that enhances foam cell formation, a crucial early event in atheroma development.

Expression of ABCA1, ABCG1 and apoE are mediated through the nuclear receptors, LXRα and peroxisome-proliferator-activated receptor (PPAR) γ [18-20]. LXRs function as sterol sensors by responding to increases in oxysterols with upregulated transcription of gene products that control cholesterol catabolism and efflux. Activation of LXRα by endogenous oxysterol ligands induces transcription of ABCA1 and ABCG1 [23-25].

PPARγ, retinoic acid receptor (RAR) and retinoid X receptor (RXR) agonists may elevate the level and activity of 27-hydroxylase, providing possible mechanisms for atheroprotection [12, 26, 27]. Overexpression of 27-hydroxylase was shown to increase efflux of cholesterol to apoA-1 via an ABCA1-independent pathway [14]. However, the mechanism underlying this phenomenon has still not been described.

Oxidized cholesterols (oxysterols) may be transported to hepatocytes and converted to bile acids for excretion from the liver. CYP27A1 is a very well described key enzyme in bile acid formation [55-57]. Wang et al. [58] demonstrated that bile extracts specifically activate the orphan receptor farnesoid X receptor (FXR) which functions as a nuclear bile acid receptor (BAR). Ligand-occupied FXR inhibits the transcriptional activity of the oxysterol receptor LXRα and positively regulates Cyp7a transcription.

Among all of the P450 enzymes, CYP27A1 and CYP2R1 are considered as prime candidates for vitamin D 25-hydroxylation [28-30]. Cyp27a1 gene knockout mice display disrupted cholesterol metabolism and bile acid synthesis plasma levels of cholesterol, retinol, tocopherol, and 1,25-dihydroxyvitamin D are unaltered [31, 32].

Similarly, patients with cerebrotendinous xanthomatosis due to mutations in the Cyp27a1 gene show reduced synthesis of bile acids, but do not display vitamin D-related pathology, likely because there are other enzymes to handle vitamin D processing [33].

In this study, we evaluate the impact of inflammatory conditions on 27-hydroxylase expression. We use the specific pro-atherogenic cytokines interferon (IFN)-γ, tumor necrosis factor (TNF)-α and interleukin (IL)-1. As diabetes is closely associated with cardiovascular disease, we also look at human serum albumin modified with N^{ε}-(carboxymethyl)lysine (CML-HSA), a major advanced glycation end-product (AGE) adduct implicated in atherosclerosis. We then test the atheroprotective potential of the stilbene resveratrol as well as the adenosine A2A receptor ligands UK-432,097 and CGS-21680. To confirm that effects are specifically mediated through the A2A receptor, we block this receptor with the selective antagonist ZM-241385.

All of these compounds are ligands to the A2A receptor and this receptor is a target of interest because our own previous work has shown that A2A agonists improve cholesterol balance [34, 35]. The drug methotrexate acts through adenosine release to reduce cardiovascular risk in inflammatory conditions such as rheumatoid arthritis (RA) [36].

We perform these studies in cell culture in macrophages and endothelial cells, cell types that are critical to the development of atherosclerosis.

Our interest then extends beyond the addition of single agents to a more accurate representation of the *in vivo* milieu which we achieve by exposing cells to plasma from patients with RA and psoriatic arthritis (PsA), rheumatologic disorders that are known to increase risk of cardiovascular events to differing degrees with greater risk seen in RA [37].

METHODS

Cell Culture

THP-1 monocytes (American Type Culture Collection, Manassas, VA) and human peripheral blood mononuclear cells (HPBMC) (Lonza Inc., Allendale, NJ) were cultured in RPMI 1640 supplemented with 10% fetal calf serum (FCS), 2mM L-glutamine and 50 µg per ml of penicillin-streptomycin at 37°C in a 5% CO_2 atmosphere to a density of 10^6 cells per ml. Cell culture media and supplementary reagents were obtained from Invitrogen (Grand Island, NY). Differentiation of the monocytic THP-1 cells into adherent macrophages was stimulated by 48h exposure to 100nM phorbol 12-myristate 13-acetate (PMA), obtained from Sigma-Aldrich (St. Louis, MO). When differentiated phenotype was achieved, the PMA-containing medium was removed and replaced with complete RPMI 1640 supplemented with 10% FCS. The macrophages and HPBMC were cultured in unadulterated medium for 24h before treatment.

Human aortic endothelial cells (HAEC) and basal medium were purchased from Lifeline Cell Technology (Walkersville, MD). HAEC were grown at 37^0C in a 5% CO_2 atmosphere in Vasculife Basal Medium supplemented with Lifeline's growth factors. HAEC were between passages 3 and 6 when the experiments were performed. When cells reached 80% confluence they were subjected to incubation under conditions described in the next section.

Experimental Conditions

The THP-1 human monocytes/macrophages were incubated (18h, 37°C, 5% CO_2, n=3 per condition) under the following conditions: untreated control (media alone), IFN-γ (500U/ml), TNF-α (100U/ml), IL-1 (10u/ml).

THP-1 macrophages and HPBMC were incubated for 30 min, 1h, 3h and 5h (37^0C, 5% CO_2) in the presence or absence of 50 µg/ml CML-HSA.

THP-1 monocytes and macrophages, HAEC, PBMC were subjected to incubation for 18h under the following experimental conditions: a) cell culture media alone; b) ethanol solvent control; c) DMSO solvent control; d) resveratrol (10µM); e) ZM-241385 (1µM, 1h) + resveratrol (10µM); f) UK-432,097 (100 nM); g) ZM-241385 (1µM) + UK-432,097 (100 nM); h) CGS-21680 (1 µM); i) ZM-241385 (1 µM, 1h) + CGS-21680 (1 µM). To obtain total protein lysates, incubation was done for 24 h under conditions a-i described above.

THP-1 differentiated macrophages were incubated (18h) in the presence of 10% plasma from the following subjects: 21 healthy controls (HC); 22 RA and 16 PsA, age and sex-matched to the HC group. The study was performed under a Winthrop University Hospital IRB-approved protocol.

Gene and Protein Expression Analysis

Following incubation, RNA and protein were isolated. QRT-PCR and Western blotting techniques were then used to measure expression of 27-hydroxylase. Each reaction was done in triplicate. Following specific primers to 27-hydroxylase were used for PCR reaction: F 5'-AAGCGATACCTGGATGGTTG-3'; R 5'-TGTTGGATGTCGTGTCCACT-3'. To correct for differences in cDNA load among samples, the target PCRs were normalized to a reference PCR involving the endogenous housekeeping gene GAPDH. Non-template controls were included for each primer pair to check for significant levels of any contaminants.

Anti-cholesterol 27-hydroxylase antibody is an affinity-purified rabbit polyclonal antibody raised against residues 15-28 of the cholesterol 27-hydoxylase protein. As a loading control, on the same transferred membrane, β-actin was detected using mouse anti-human β-actin antibody (ab8227, Abcam Inc., Cambrige, MA).

Statistical Analysis of Experimental Data

Statistical analysis was performed using Graphpad Prism, version 5.01. Pairwise multiple comparison was made between control and treatment conditions using unpaired t tests, two tailed 95% confidence intervals, significance $p < 0.05$.

RESULTS AND DISCUSSION

CYP27A1 expression is substantially up-regulated in atherosclerotic lesions and its abundance increases with lesion progression [38, 39]. Genetic heterogeneity resulting in differences in cholesterol 27-hydroxylase activity among individuals may affect their ability to deal with accumulated cholesterol in the arterial intima [40].

Cholesterol loading dose-dependently stimulates the production of 27-hydroxycholesterol and induction of ABCA1 and ABCG1 expression in human macrophages [41]. CYP27A1 expression increases during monocyte-to-macrophage differentiation [42]. Cholesterol efflux and CYP27A1 expression are suppressed by some inflammatory mediators (IFN-γ or immune complexes, but not IL-1 or TNF-α) in endothelial cells and monocytes/macrophages [43] (Figure 2).

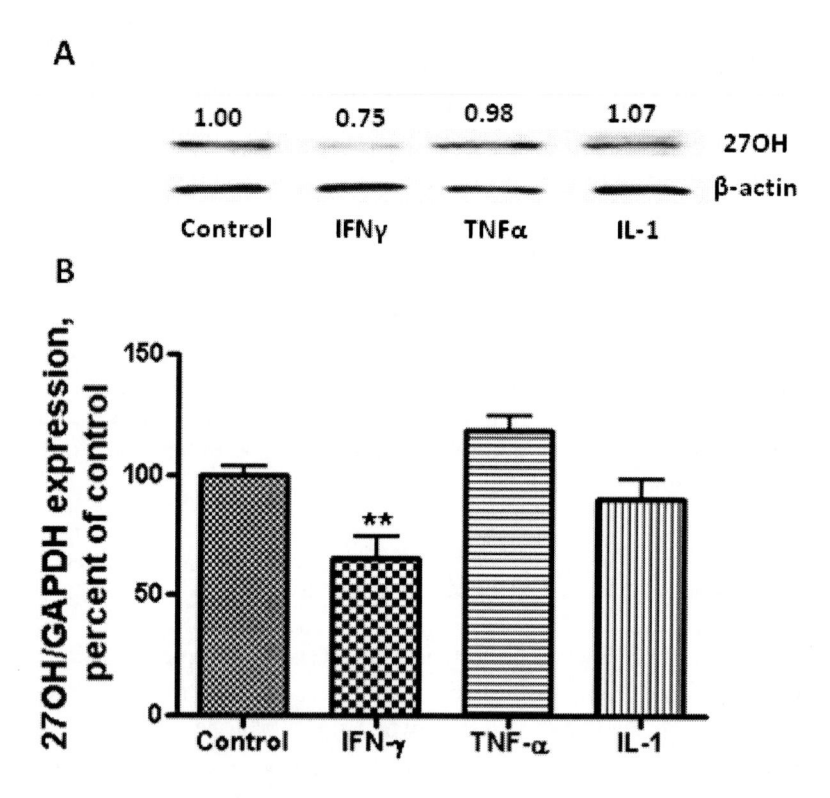

Figure 2. The effect of pro-inflammatory cytokines on 27-hydroxylase (27-OH) message and protein levels in THP-1 macrophages. Only IFN-γ significantly downregulates protein (A) and mRNA (B) expression of 27-hydroxylase. All results are presented as means ± SEM of three independent experiments. ** - P<0.01 vs. control cells.

We recently demonstrated that AGEs, glycated proteins or lipids that contribute to atherosclerosis, decrease message and protein level of 27-

hydroxylase (by 54.5±2.9% and 48.7±9.23%, respectively [at 5h]) in THP-1 human macrophages (Figure 3). The effect of the AGE CML-HSA on expression of 27-hydroxylase in HPBMC was less pronounced, reducing the mRNA level by 37.2±3.5% (n=3, P<0.05) (Figure 3).

Nunes et al. [44] demonstrated that the plasma ratio of 27-hydroxycholesterol/total cholesterol (median and range nmoL/mmoL) is higher in healthy subjects with low high density lipoprotein (HDL) cholesterol (p=0.0258). Their data indicate that the production of 27-hydroxycholesterol by extrahepatic tissues and its transport to the liver may represent an alternative pathway for reverse cholesterol transport when the system is compromised by low plasma HDL.

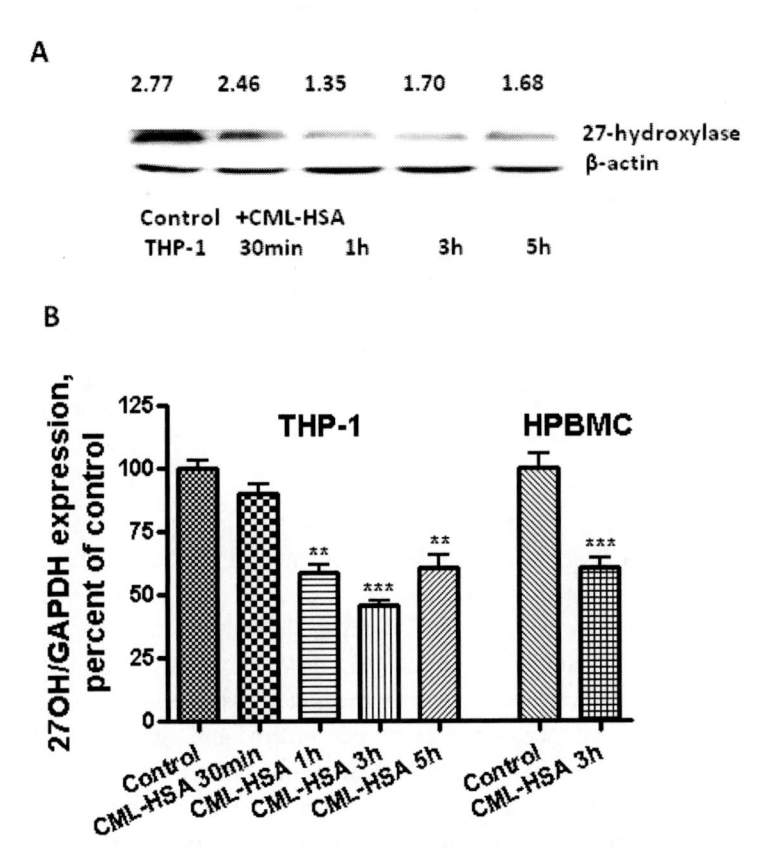

Figure 3. CML-HSA represses expression of 27-hydroxylase in THP-1 human macrophages and HPBMC. Incubation with CML-HSA significantly decreases 27-hydroxylase protein (**A**) and message (**B**) over time. All results are presented as means ± SEM of three independent experiments. ** - P<0.01; *** - P<0.001 vs. control cells.

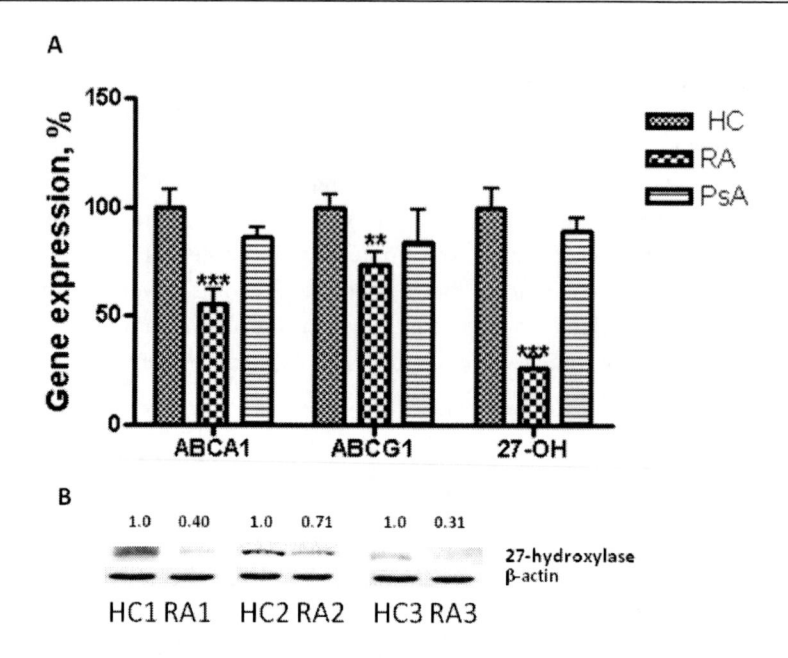

Figure 4. Effect of RA and PsA plasma on the expression of cholesterol efflux proteins in THP-1 human macrophages. THP-1 macrophages were incubated in the presence of 10% plasma from each of RA, PsA or HC subjects for 18h. (**A**) Gene expression levels were graphed as relative mRNA expression with mean of HC set at 100%. The data represent the mean for 22 RA, 16 PsA or 21 HC ± SEM of three independent experiments. ** - P<0.01; *** - P<0.001 versus THP-1 macrophages exposed to HC plasma. (**B**) Protein expression presented as protein abundance for three HC and three RA patients.

Two autoimmune rheumatic disorders, systemic lupus erythematosus (SLE) and RA, are associated with high incidence of cardiovascular disease. Atherosclerosis contributes to approximately 30% of deaths in patients with SLE and cardiovascular mortality is 50% higher in patients with RA compared with age- and sex-matched controls [45, 46]. Previously published data indicates that plasma from patients with rheumatic diseases such as SLE and RA disrupts the balance between lipid efflux and influx in human macrophages compared to plasma from healthy individuals (HC). Exposure of macrophages and endothelial cells to plasma from patients with SLE suppresses 27-hydroxylase [47] and increases foam cell transformation in THP-1 macrophages. The average expression of 27-hydroxylase in THP-1 macrophages in the presence of RA plasma diminishes to 26.5±8.56% (P<0.01) for mRNA and to 40.12±10.23% (P<0.001) for protein [48]. RA plasma exhibits elevation of inflammatory markers, notably IFN-γ and also greater atherogenicity than plasma from a milder rheumatic disease, PsA (Figure 4). A more modest inflammatory pattern in PsA and no significant effects on cholesterol transport gene expression are consistent with less cardiovascular disease risk in PsA compared to RA.

Adenosine, a purine nucleoside and critical modulator of inflammation, has been shown previously to have atheroprotective properties mediated through activation of the adenosine A2A receptor. Our group has demonstrated that A2A receptor ligation reduces the development of foam cells and augments the expression of the ABCA1 and 27-hydroxylase proteins [34, 35]. Methotrexate, a disease modifying anti-rheumatic drug used to treat RA, acts on cells to promote adenosine release [36]. It has long been known that methotrexate reduces cardiovascular events in RA patients and this may be due to adenosine effects not only on inflammation, but also on cholesterol metabolism [49].

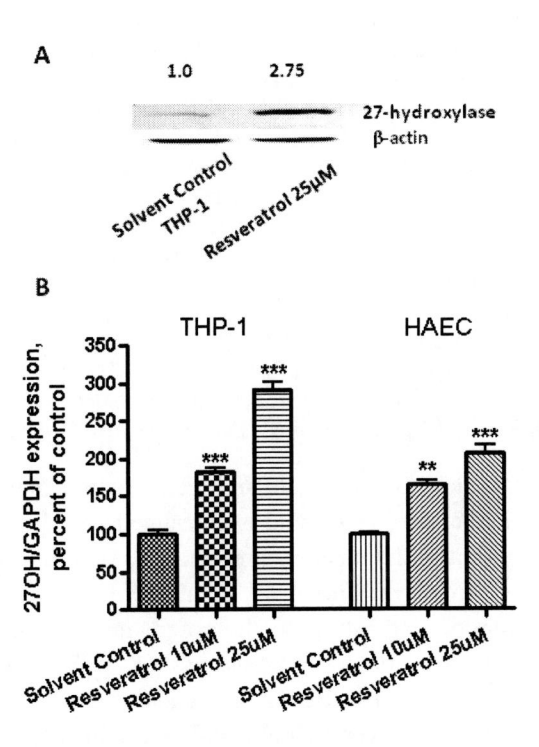

Figure 5. Effect of resveratrol on the expression of 27-hydroxylase (27-OH) in THP-1 human macrophages and HAEC. Resveratrol concentration dependently augments mRNA and protein expression of 27-hydroxylase in cultured THP-1 macrophages and HAEC. ** - P<0.01; *** - P<0.001 versus solvent control set at 100%.

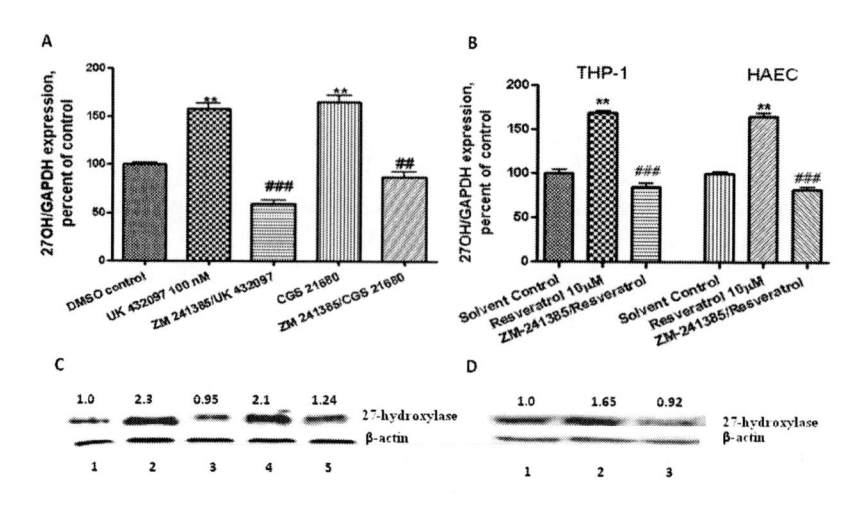

Figure 6. Effect of adenosine ligation on the expression of 27-hydroxylase (27-OH) in THP-1 human macrophages and HAEC. (A) Relative 27-hydroxylase mRNA abundance increased upon exposure to UK432,097 (100 nM) and CGS-21680 (1 μM). This effect was abolished upon pre-incubation with the adenosine 2A receptor antagonist ZM-241385 (1 μM). (**B**). Salutary effect of resveratrol on the expression of 27-hydroxylase is similarly reduced by pre-incubation with the adenosine 2A receptor antagonist ZM-241385 (1 μM) in THP-1 macrophages and HAEC. The data represent the mean and SEM of three independent experiments (n=3). * - P<0.05, ** - P<0.01 vs. solvent control, # - P<0.05, ### - P<0.001 vs. A2AR agonists. Immunoblot results presented as protein abundance of 27-hydroxylase in THP-1 macrophages with normalization to β-actin. (**C**) THP-1 macrophages were incubated under following conditions: 1 - DMSO solvent control; 2 - UK-432,097 (100 nM); 3 - ZM-241385 (1μM) + UK-432,097 (100 nM); 4 - CGS-21680 (1μM); 5 - ZM-241385 (1 μM, 1h) + CGS-21680 (1 μM). (**D**) THP-1 macrophages were subjected to: 1 - ethanol solvent control; 2 - resveratrol (10μM); 3 - ZM-241385 (1μM, 1h) + resveratrol (10μM).

Resveratrol is a bioactive molecule and dietary supplement found in red wine, grapes and herbal medicines, and is consumed worldwide. The known cardioprotective and anti-inflammatory properties of resveratrol have spurred investigation of the mechanisms involved. A number of previous studies have indicated that resveratrol can influence cholesterol metabolism [50, 51]. Our group revealed that resveratrol affects cholesterol efflux by enhancing reverse cholesterol transport (ABCA1, ABCG1 and scavenger receptor-B1) and by alterations in cholesterol 27-hydroxylase expression and function [52]. In our study, resveratrol treatment increases levels of 24s- and 7α-hydroxycholesterols, 24- and 7-oxocholesterols in THP-1 macrophages (Figure 5). In addition to 27-hydroxycholesterol, these oxysterols are weak endogenous ligands for LXRα and our observation could present another mechanism of LXRα-pathway activation by resveratrol.

Resveratrol causes an increase in plasma adenosine levels and blood nucleosides in human subjects [53, 54]. In our study, pharmacologic blockade of the A2AR with ZM-241385 eliminated beneficial effects of resveratrol on cholesterol efflux. Thus, an adenosinergic mechanism may play a role in resveratrol effects on cholesterol metabolism.

We observed stimulation in 27-hydroxylase expression in human THP-1 macrophages and HPBMC upon A2A receptor ligation (CGS-21680, UK-432,097, reseveratrol, methotrexate). This effect was abolished by the selective A2A receptor antagonist ZM-241385 (Figure 6).

CONCLUSION AND PROSPECSTIVES

Multiple mechanisms for removal of cholesterol involve P450 enzymes. Egress of cholesterol from extrahepatic tissues involves conversion of cholesterol into the oxysterol 27-hydroxycholesterol by the P450 27-hydroxylase. 27-Hydroxycholesterol and other 27-oxygenated steroids are more polar than cholesterol and can be transported from macrophages and other cells more efficiently than cholesterol. 27-Hydroxycholesterol suppresses the rate-limiting enzyme in cholesterol biosynthesis, HMGCoA reductase. The 27-oxygenated steroid products of 27-hydroxylase catalytic activity are taken up and metabolized by the liver for excretion in the bile. Moreover, they function as ligands for LXR, activating a pathway that enhances expression of proteins that promote cholesterol efflux such as ABCA1 and ABCG1.

The enzymatic step considered to be the most important for the regulation of bile acid formation is the 7α-hydroxylation of cholesterol by the liver-specific enzyme cholesterol 7α-hydroxylase (CYP7A). CYP7A1 activity correlates with lower cholesterol levels, and vice versa. Deficiency in CYP7A1 causes a decrease in bile acid production and manifests with marked hypercholesterolemia as well as high LDL, susceptibility to gallstones, premature coronary and peripheral vascular disease.

Our lab is focused on unraveling the interaction of the pro-inflammatory environment with cells of the vessel wall that lead to accelerated atherosclerosis. Disruption of cholesterol balance is fostered under inflammatory conditions. We have documented suppression of 27-hydroxylase activity in macrophages and endothelium upon exposure to inflammatory mediators associated with autoimmune disorders and diabetes. Anti-inflammatory agents that act through the adenosine A2A receptor induce 27-hydroxylase expression, a mechanism that may account for atheroprotective properties of the drug methotrexate. New and improved drugs and dietary supplements that activate the A2A pathway may reduce cardiovascular risk in SLE, RA and diabetes. The Cardiovascular Inflammation Reduction Trial (CIRT) is an ongoing randomized, double blind, placebo-controlled clinical trial designed to determine whether low dose methotrexate will reduce recurrent cardiovascular events in 7,000 stable post-myocardial infarction patients with type 2 diabetes or metabolic syndrome [55].

REFERENCES

[1]   Nebert D.W., Russell D.W., (2002). Clinical importance of the cytochromes P450. *Lancet.* 360(9340), 1155-1162.

[2]   Orellana M., Guajardo V., (2004). Cytochrome P450 activity and its alteration in different diseases. *Rev. Med. Chil.*132, 85–94.

[3]   Luoma P.V., Pelkonen R.O., Sotaniemi E.A., (1979). Plasma high density lipoprotein cholesterol and hepatic drug metabolizing enzyme activity in man. *Acta Physiol Scand Suppl* 473:71.

[4]   Luoma P.V., Sotaniemi E.A., Pelkonen R.O., (1983). Inverse relation of serum LDL cholesterol and the LDL/HDL cholesterol ratio to liver microsomal induction in man. *Res Commun Chem Pathol Pharmacol* 42, 173–176.

[5]   Luoma P.V., (1997). Gene activation, apolipoprotein A-I/ high density lipoprotein, atherosclerosis prevention and longevity. *Pharmacol Toxicol* 81, 57–64.

[6]   Bjorkhem I., Andersson O., Diczfalusy U., Sevastik B., Xiu R.J., Duan C., Lund E., (1994). Atherosclerosis and sterol 27-hydroxylase: evidence for a role of this enzyme in elimination of cholesterol from human macrophages. *Proc Natl Acad Sci USA.* 91, 8592 – 8596.

[7]   Babiker A., Andersson O., Lund E., Xiu R.J., Deeb S., Reshef A., Leitersdorf E., Diczfalusy U., Björkhem I., (1997). Elimination of cholesterol in macrophages and endothelial cells by the sterol 27-hydroxylase mechanism. *J. Biol. Chem.* 272, 26253-26261.

[8]   Reiss A.B., Awadallah N.W., Cronstein B.N., (2000). Cytochrome P450 cholesterol 27-hydroxylase: an anti-atherogenic enzyme. *Recent Research Devel in Lipids Res.* 4, 39-50.

[9] Luoma P.V., (2008). Cytochrome P450 and gene activation--from pharmacology to cholesterol elimination and regression of atherosclerosis. *Eur J Clin Pharmacol.* 64(9), 841-850.

[10] Schroepfer G.J. Jr., (2000). Oxysterols: modulators of cholesterol metabolism and other processes. *Physiol Rev.* 80, 361–554.

[11] Shanahan C.M., Carpenter K.L., Cary N.R., (2001). A potential role for sterol 27-hydroxylase in atherosclerosis. *Atherosclerosis* 154, 269 – 276.

[12] Escher G., Krozowski Z., Croft K.D., Sviridov D., (2003). Expression of sterol 27-hydroxylase (CYP27A1) enhances cholesterol efflux. *J. Biol. Chem.* 278, 11015–11019.

[13] Cali J.J., Hsieh C.L., Francke U., Russell D.W., (1991). Mutations in the bile acid biosynthetic enzyme sterol 27-hydroxylase underlie cerebrotendinous xanthomatosis. *J. Biol. Chem.* 266, 7779–7783.

[14] Lund E., Andersson O., Zhang J., Babiker A., Ahlborg G., Diczfalusy U., Einarsson K., Sjovall J., Bjorkhem I. (1996). Importance of a novel oxidative mechanism for elimination of intracellular cholesterol in humans. *Arterioscler. Thromb. Vasc. Biol.* 16, 208–212.

[15] Lehmann J.M., Kliewer S.A., Moore L.B., Smith-Oliver T.A., Oliver B.B., Su J.L., Sundseth S.S., Winegar D.A., Blanchard D.E., Spencer T.A., (1997). Activation of the nuclear receptor LXR by oxysterols defines a new hormone response pathway. *J. Biol. Chem.* 272: 3137–3140.

[16] Janowski, B.A., Grogan M.J., Jones S.A., Wisely G.B., Kliewer S.A., Corey E.J., Mangelsdorf D.J., (1999). Structural requirements of ligands for the oxysterol liver X receptors LXRalpha and LXRbeta. *Proc. Natl. Acad. Sci. USA.* 96, 266–271.

[17] Fu X., Menke J.G., Chen Y., Zhou G., MacNaul K.L., Wright S.D.,. Sparrow C.P, Lund E.G., (2001). 27-hydroxycholesterol is an endogenous ligand for liver X receptor in cholesterol-loaded cells. *J Biol Chem* 276, 38378-38387.

[18] Venkateswaran A., Laffitte B.A, Joseph S.B., Mak P.A., Wilpitz D.C., Edwards P.A., Tontonoz P., (2000). Control of cellular cholesterol efflux by the nuclear oxysterol receptor LXR alpha. *Proc. Natl. Acad. Sci. USA.* 97, 12097–12102

[19] Tabas I., (2002). Consequences of cellular cholesterol accumulation: basic concepts and physiological implications. *J Clin Invest.* 110, 905–911.

[20] Voloshyna I., Reiss A.B., (2011). The ABC transporters in lipid flux and atherosclerosis. *Prog Lipid Res.* 50(3), 213-224.

[21] Oram, J.F., Lawn R.M., (20010 ABCA1: the gatekeeper for eliminating excess tissue cholesterol. *J. Lipid Res.* 42, 1173–1179.

[22] Zanotti I., Pedrelli M., Potì F., Stomeo G., Gomaraschi M., Calabresi L., Bernini F., (2011). Macrophage, but not systemic, apolipoprotein E is necessary for macrophage reverse cholesterol transport in vivo. *Arterioscler Thromb Vasc Biol* 31(1), 74-80.

[23] Szanto A., Benko S., Szatmari I., Balint B. L, Furtos I., Ruhl R., Molnar S., Csiba L., Garuti R., Calandra S., (2004). Transcriptional regulation of human CYP27 integrates retinoid, peroxisome proliferator-activated receptor, and liver X receptor signaling in macrophages. *Mol. Cell. Biol.* 24, 8154–8166.

[24] Zelcer N., Tontonoz P., (2006). Liver X receptors as integrators of metabolic and inflammatory signaling. *J. Clin. Invest.*116(3), 607-614.

[25] Ricote M., Valledor A.F., Glass C.K., (2004). Decoding transcriptional programs regulated by PPARs and LXRs in the macrophage: effects on lipid homeostasis, inflammation, and atherosclerosis. *Arterioscler Thromb Vasc Biol.* 24, 230–239.

[26] Quinn C.M., Jessup W., Wong J., Kritharides L., Brown A.J., (2005). Expression and regulation of sterol 27-hydroxylase (CYP27A1) in human macrophages: a role for RXR and PPARgamma ligands. *Biochem J.* 385(Pt 3), 823-830.

[27] Li T., Chen W., Chiang J.Y., (2007). PXR induces CYP27A1 and regulates cholesterol metabolism in the intestine. *J Lipid Res.* 48(2), 373-384.

[28] Cali J.J., Russell D.W., (1991). Characterization of human sterol 27-hydroxylase. A mitochondrial cytochrome P-450 that catalyzes multiple oxidation reaction in bile acid biosynthesis. *J Biol Chem* 266(12), 7774–7778

[29] Norlin M., Wikvall K., (2007). Enzymes in the conversion of cholesterol into bile acids. *Curr Mol Med* 7(2), 199–218.

[30] Lorbek G., Lewinska M., Rozman D., (2012). Cytochrome P450s in the synthesis of cholesterol and bile acids—from mouse models to human diseases. *FEBS J* 279(9), 1516–1533.

[31] Rosen H., Reshef A., Maeda N., Lippoldt A., Shpizen S., Triger L., Eggertsen G., Björkhem I., Leitersdorf E., (1998). Markedly reduced bile acid synthesis but maintained levels of cholesterol and vitamin D metabolites in mice with disrupted sterol 27-hydroxylase gene. *J Biol Chem.* 273(24), 14805-14812.

[32] Repa J.J., Lund E.G., Horton J.D., Leitersdorf E., Russell D.W., Dietschy J.M., Turley S.D., (2000). Disruption of the sterol 27-hydroxylase gene in mice results in hepatomegaly and hypertriglyceridemia. Reversal by cholic acid feeding. *J Biol Chem* 275(50), 39685–39692.

[33] Gallus G.N., Dotti M.T., Federico A., (2006). Clinical and molecular diagnosis of cerebrotendinous xanthomatosis with a review of the mutations in the CYP27A1 gene. *Neurol Sci* 27(2),143–149.

[34] Reiss A.B., Rahman M.M., Chan E.S., Montesinos M.C., Awadallah N.W., Cronstein B.N., (2004). Adenosine A2A receptor occupancy stimulates expression of proteins involved in reverse cholesterol transport and inhibits foam cell formation in macrophages. *J Leukoc Biol.* 76(3), 727-734.

[35] Voloshyna I., Carsons S., Littlefield M.J., Rieger J.M., Figler R., Reiss A.B., (2013). Adenosine A(2A) receptor activation supports an atheroprotective cholesterol balance in human macrophages and endothelial cells. *Biochim. Biophys. Acta.* 1831(2), 407-416.

[36] Riksen N.P., Barrera P., van den Broek P.H., van Riel P.L., Smits P., Rongen G.A., (2006). Methotrexate modulates the kinetics of adenosine in humans in vivo. *Ann Rheum Dis* 65(4), 465-470.

[37] Crowson C.S., Matteson E.L., Roger V.L., Therneau T.M., Gabriel S.E., (2012). Usefulness of risk scores to estimate the risk of cardiovascular disease in patients with rheumatoid arthritis. *Am J Cardiol* 110, 420–424.

[38] Bjorkhem I., Andersson O., Diczfalusy U., Sevastik B., Xiu R. J., Duan C., Lund E., (1994). Atherosclerosis and sterol 27-hydroxylase: evidence for a role of this enzyme in elimination of cholesterol from human macrophages. *Proc. Natl. Acad. Sci. USA.* 91,8592–8596.

[39] Crisby M., Nilsson J., Kostulas V., Bjorkhem I., Diczfalusy U., (1997). Localization of sterol 27-hydroxylase immuno-reactivity in human atherosclerotic plaques. *Biochim. Biophys. Acta.* 1344, 278–285.

[40] Shanahan C.M., Carpenter K.L., Cary N.R., (2001). A potential role for sterol 27- hydroxylase in atherosclerosis. *Atherosclerosis* 154, 269 – 276.

[41] von Bahr S., Movin T., Papadogiannakis N., Pikuleva I., Rönnow P., Diczfalusy U., Björkhem I., (2002). Mechanism of accumulation of cholesterol and cholestanol in tendons and the role of sterol 27-hydroxylase (CYP27A1). *Arterioscler Thromb Vasc Biol.* 22(7),1129-1135.

[42] Hansson M., Ellis E., Hunt M. C., Schmitz G., Babiker A., (2003). Marked induction of sterol 27-hydroxylase activity and mRNA levels during differentiation of human cultured monocytes into macrophages. *Biochim. Biophys. Acta.* 1593, 283–289.

[43] Reiss A.B., Awadallah N.W., Malhotra S., Montesinos M.C., Chan E.S., Javitt N.B., Cronstein B.N., (2001). Immune complexes and IFN-gamma decrease cholesterol 27-hydroxylase in human arterial endothelium and macrophages. *J Lipid Res.* 42(11), 1913-1922.

[44] Nunes V.S., Leança C.C., Panzoldo N.B., Parra E., Zago V., Cazita P.M., Nakandakare E.R., de Faria E.C., Quintão E.C., (2013). Plasma 27-hydroxycholesterol/cholesterol ratio is increased in low high density lipoprotein-cholesterol healthy subjects. *Clin Biochem.* 46(15), 1619-1621.

[45] Sandoo A., Kitas G.D., Carroll D., Veldhuijzen van Zanten J.J., (2012). The role of inflammation and cardiovascular disease risk on microvascular and macrovascular endothelial function in patients with rheumatoid arthritis: a cross-sectional and longitudinal study. *Arthritis Res Ther* 14, R117.

[46] Crowson C.S., Matteson E.L., Roger V.L., Therneau T.M., Gabriel S.E., (2012). Usefulness of risk scores to estimate the risk of cardiovascular disease in patients with rheumatoid arthritis. *Am J Cardiol* 110, 420–424.

[47] Reiss A.B., Anwar K., Merrill J.T., Chan E.S., Awadallah N.W., Cronstein B.N., Michael Belmont H., Belilos E., Rosenblum G., Belostocki K., Bonetti L., Hasneen K., Carsons S.E., (2010). Plasma from systemic lupus patients compromises cholesterol homeostasis: a potential mechanism linking autoimmunity to atherosclerotic cardiovascular disease. *Rheumatol Int.* 30(5), 591-598.

[48] Voloshyna I., Modayil S., Littlefield M.J., Belilos E., Belostocki K., Bonetti L., Rosenblum G., Carsons S.E., Reiss A.B., (2013). Plasma from rheumatoid arthritis patients promotes pro-atherogenic cholesterol transport gene expression in THP-1 human macrophages. *Exp Biol Med (Maywood).* 238(10), 1192-1197.

[49] van Halm V., Nurmohamed M.T., Twisk J.W., Dijkmans B.A., Voskuyl A.E., (2006). Disease-modifying antirheumatic drugs are associated with a reduced risk for cardiovascular disease in patients with rheumatoid arthritis: a case control study. *Arthritis Research Therapy*, 8(5), R151.

[50] Sevov M., Elfineh L., Cavelier L.B., (2006). Resveratrol regulates the expression of LXR-alpha in human macrophages. *Biochem. Biophys. Res. Commun.* 348(3),1047-1054.

[51] Shah P.K., Patel J.A., (2010). Resveratrol and its biological actions. *Intl. J. Green. Pharmacy.* 4(1), 15-21.

[52] Voloshyna I., Hai O., Littlefield M.J., Carsons S., Reiss A.B., (2013). Resveratrol mediates anti-atherogenic effects on cholesterol flux in human macrophages and endothelium via PPARγ and adenosine. *Eur J Pharmacol.* 698(1-3), 299-309.

[53] Bradamante S., Piccinini F., Barenghi L., Bertelli A.A., De Jonge R., Beemster P., De Jong J.W., (2000). Does resveratrol induce pharmacological preconditioning? *Int. J. Tissue React.* 22(1), 1-4.

[54] Imamura G., Bertelli A.A., Bertelli A., Otani H., Maulik N., Das D.K., (2002). Pharmacological preconditioning with resveratrol: an insight with iNOS knockout mice. *Am. J. Physiol. Heart. Circ. Physiol.* 282, H1996–H2003.

[55] Ridker P.M., (2009). Testing the inflammatory hypothesis of atherothrombosis: scientific rationale for the cardiovascular inflammation reduction trial (CIRT). *J Thromb Haemost.* 7 Suppl 1, 332-339.

In: Cytochrome P450 Enzymes
Editor: Jian Wu

ISBN: 978-1-61942-209-4
© 2014 Nova Science Publishers, Inc.

Chapter 5

DRUG RESISTANCE OF NUCLEOS(T)IDE ANALOGUES AND RELATED ADVERSE EVENTS IN CHRONIC HEPATITIS B

Li Zong[*]

School of Basic Medical Sciences, Fudan University, Shanghai, China

ABSTRACT

Chronic hepatitis B is a major health burden and a cause of liver cirrhosis as well as hepatocellular carcinoma. Nucleos(t)ide analogues are effective in controlling the disease and decreasing the occurrence of hepatocellular carcinoma, but drug resistance will limit their long-time effectiveness. What's more, nucleos(t)ide analogues could also cause serious adverse events. Understanding the mechanisms of drug resistance and adverse events will enable researchers to design new agents and devise strategies to minimize the development of antiviral drug resistance or adverse events. This paper will review mechanisms of drug resistance and adverse events related to nucleos(t)ide analogues.

Keywords: Nucleos(t)ide analogues, chronic hepatitis B, drug resistance, adverse events

---

[*] Address: Shanghai Medical College, Fudan University, Shanghai 200032, China; E-mail:zongli1226@163.com

Chronic hepatitis B (CHB) is a major health burden, with an estimated 400 million people affected globally. Up to 40% of those with CHB may develop complications, including cirrhosis, decompensated liver disease and hepatocellular carcinoma (HCC).[1] The importance of HBV DNA levels in determining long-term outcomes in CHB patients has been demonstrated in several large prospective studies. In a study of 2763 CHB patients, their relative risk of mortality from HCC was 11.2 in patients with levels of $1.6 \times 103-105$ copies/mL. In the REVEAL study of 3653 CHB patients, HBV DNA levels $\geqq 2000$ IU/mL were associated with significantly higher risk of HCC.[2] This same cohort also showed that the level of HBV DNA was an independent risk factor for the development of cirrhosis. The levels of HBV DNA have been shown to correlate well with liver histological activity and with fibrosis stages. HBV can cause HCC partly via integration of its DNA into the host genome but more importantly by inciting chronic injury to the hepatocytes, leading to an increase in cellular turn over. [3]

HBV GENOME AND REPLICATION

HBV is a small DNA virus. The infectious virion circulates as a 42-nm Dane particle that comprises a nucleocapsid surrounded by a lipid envelope. The nucleocapsid contains a partially double-stranded genome of ~3.2-kilobase. The compact genome encodes 4 overlapping reading frames which through 4 RNA species generate 7 viral proteins: large, middle, and small envelope proteins that form the HBsAg, the nucleocapsid core protein, the secretory hepatitis B e antigen, the viral reverse transcriptase/polymerase, and the X protein. [4]

Most studies of HBV have shown that the Pre-S1 domain of the large envelope protein is required for receptor binding and initiation of infection. Once the virus is inside a susceptible cell, viral nucleocapsids disassemble, and genomic DNA is transferred to the nucleus. Inside the nucleus, the partially double-stranded viral DNA is converted to cccDNA. The cccDNA becomes a viral minichromosome, acting as the major transcriptional template for the virus. [5] The viral mRNAs are transported to the cytoplasm where translation of viral proteins, nucleocapsid assembly, and viral replication occurs. Replication occurs within a nucleocapsid that consists of the core protein, the pregenomic RNA, and the polymerase. Nucleocapsid formation requires the coordinated binding of the polymerase to an RNA stem-loop structure at the 5'end of the pregenomic RNA called epsilon, which triggers encapsidation. The polymerase bound to epsilon serves as a protein primer for DNA synthesis with epsilon serving as the template for this reaction. After completion of the negative-strand DNA synthesis, the RNA is degraded by the RNase H activity of the polymerase, followed by positive-strand DNA synthesis and circularization of the viral genome. Viral nucleocapsid with (partially) double stranded DNA interacts with the envelope proteins in the endoplasmic reticulum to form mature virions that are secreted from the cell. Viral nucleocapsids can also be transported back to the nucleus to maintain a pool of cccDNA.

NUCLEOS(T)IDE ANALOGUES

HBV polymerase is a multifunctional protein with RNA- and DNA-dependent DNA polymerase functions that are essential for viral replication. It contains four domains: the terminal protein (important for initiating HBV replication and nucleocapsid packaging), the spacer region, the reverse transcriptase and the RNase H domain (important for degradation of pregenomic RNA template). [6] Nucleos(t)ide analogues have similar structures to the natural nucleotides and compete at the HBV polymerase catalytic site during the synthesis of viral DNA. They lack a hydroxyl group, preventing the formation of a covalent bond with the adjoining nucleotide, causing chain termination. Three categories of agents are currently available: L-nucleosides, acyclic phosphonate nucleotides, and cyclopentane deoxyguanosine analogues. The three nucleoside analogues are lamivudine, telbivudine and entecavir, while the two nucleotide analogues are adefovir and tenofovir. [7]

DEFINITION OF VIROLOGICAL BREAKTHROUGH (VBT), GENOTYPIC RESISTANCE (GR), AND BIOCHEMICAL BREAKTHROUGH (BBT)

VBT is defined as any increase in serum HBV DNA by >1 log10 from nadir (lowest-ever) or redetection of serum HBV DNA at levels ≧10-fold the lower limit of detection of the HBV DNA assay after having an undetectable result. Thus, a patient who previously had undetectable serum HBV DNA by an assay with a lower limit of detection of 29 IU/mL would be considered to have a VBT if serum HBV DNA is subsequently detected at levels $\geqq290$ IU/mL. A cutoff $\geqq10$-fold the lower limit of detection was chosen because the consensus definition of VBT requires a 10-fold increase in HBV DNA. GR is defined as detection of signature resistance mutations by direct sequencing. BBT is defined as serum alanine aminotransferase (ALT) above the upper limit of normal (ULN) (35 IU/L) in patients who had normalized ALT and ALT >2 times nadir in those who never had normal ALT. ALT flare is defined as ALT>5 times ULN in patients who had normalized ALT and ALT>5 times nadir in those who never had normal ALT. [8]

MOLECULAR MECHANISMS OF RESISTANCE TO NAS

HBV has a high rate of replication, with 1012 virions produced per day and a high mutational rate of approximately 105 substitution/base/cycle. [9] This translates to approximately 1010 –11point mutations produced per day in individuals with active replication. Because the HBV genome is only$\geqq3200$ base pairs, all possible single base changes can be produced per day. The HBV reverse transcriptase does not have a proofreading function to repair incorrectly incorporated nucleotides. Therefore, mutations can arise very rapidly. Prior to therapy, a diverse swarm of viruses (quasispecies), including mutants with single and double mutations potentially associated with drug resistance, probably exists. The probability of a mutation being selected during therapy depends on the ability of a drug to suppress viral replication. Hence, a drug with low antiviral activity does not exert substantial selection pressure on the virus, and the chance of drug resistance is not high. Conversely, complete suppression of viral replication allows little opportunity for resistance to emerge because mutagenesis is replication dependent. NAs inhibit viral replication but do not eliminate existing virus or affect cccDNA in any major way. Monotherapy exerting modest antiviral activity and directed at 1 single target site would result in the highest probability of selecting drug resistance. The ideal treatment regimen should have antiviral activities targeted at different sites to reduce the risk of selecting out drug-resistant species. Resistance emerges when replication occurs in the presence of the drug selection pressure. Therefore, if we could achieve a complete suppression of replication, resistance would not be an issue. Other factors contributing to the emergence of drug resistance are genetic barriers to the development of mutations, mechanism of drug resistance, viral replication space, and various host factors involved in controlling viral replication. [10] Several broad principles on the mechanism of resistance have emerged from studies. First, most of the primary mutations cluster in the vicinity of the incoming nucleotide and act by directly affecting the position or stability of the bound substrate, template, or primer. In contrast, secondary mutations tend to occur away from the nucleotide-binding pocket. Finally, the structural and functional consequences of mutations are quite variable among the different agents. [10]

Lamivudine

Lamivudine [(-)28-deoxy-38-thiacytidine, 3TC] is a member of a class of antiviral nucleoside analogs that inhibit hepadnavirus replication specifically by terminating viral DNA synthesis. It was approved in 1998 for the treatment of CHB. Lamivudine resistance has been mapped to mutations in the tyrosine-methionine-aspartate-aspartate (YMDD) motif of the reverse transcriptase (rt) domain of HBV DNA polymerase. Variants emerging during lamivudine treatment show mutations in the viral polymerase within the catalytic domain (C domain), which includes the YMDD motif (for example, M204V or M204I), and within the B domain (such as L180M or V173L). [11 12] These mutants have a reduced replicative capacity compared with the wild type virus. The most common mutation is the substitution of methionine to isoleucine or valine (rtM204V/I) at the highly conserved YMDD motif of the reverse transcriptase. Four major patterns have been observed: L18OM+M204V; M204I; L180M+M204I; V173L+L180M+M204V; and occasionally L180M+M204V/I. The L18OM+M204V occurs most often. The incidence of lamivudine resistance is 15–20% per year, with 70% of patients becoming resistant after five years of treatment. [13]

Adefovir

Adefovir dipivoxil is the second nucleotide analogue approved (in 2002) for treating CHB. Adefovir dipivoxil is a guanosine analogue with a flexible acyclic linker instead of the unnatural L-nucleoside ring of lamivudine. Development of resistance to adefovir is slower compared with lamivudine, with the reported rate being 2% at 2 years and 29% at 5 years.14 It displays potent anti-HBV activity similar to lamivudine. In pivotal studies in both hepatitis B e antigen (HBeAg)-positive and HBeAg-negative patients with CHB and in studies enrolling patients with lamivudine-resistant HBV, adefovir dipivoxil has been shown to significantly reduce HBV DNA levels in most treated patients and to produce improvement in liver histology and ALT levels. One of the advantageous properties of adefovir is its efficacy against lamivudine-resistant mutants. [15, 16] HBV harboring the rtM204V/I mutations remain susceptible to adefovir. The rtA181V/T and rtN236T are the two major mutations of the HBV polymerase gene responsible for adefovir resistance. The rtN236T mutant is sensitive to lamivudine, while the rtA181T mutant is less sensitive to lamivudine. Another mutation, rtI233V, has been associated with primary adefovir resistance. However, nucleotide phosphonates adefovir and tenofovir are not cross-resistant to M204 YMDD mutant HBV. Nevertheless, adefovir resistance is more likely to emerge in patients with lamivudine resistant HBV, at rates of 18% and 25% after 1 and 2 years, respectively. [17, 18]

Entecavir

In 2005, entecavir was approved at a dose of 0.5 mg/day and 1 mg/day for treatment of treatment-naive and lamivudine-resistant CHB, respectively. Compared with lamivudine, entecavir demonstrated a 30- to 2200-fold increase in reducing viral DNA replication in vitro. Entecavir affects multiple functions of the polymerase, including priming, reverse transcription, and DNA elongation. Entecavir has a high barrier to resistance. Therefore the development of drug resistance is rare, occurring at a rate of only 1.2% after 5 years of therapy in treatment-naive patients. The reason for the high genetic barrier of entecavir is because it requires a combination of three mutations before resistance develops. [19] In addition to the rtM204V and rtL180M mutations that are responsible for resistance to lamivudine, an additional mutation at rtI169T, rtT184G, rtS202I or rtM250V is required for entecavir resistance. [20] The presence of lamivudine resistance mutations thus increases the risk of developing entecavir resistance. In lamivudine-refractory patients, the resistance rate to entecavir is 51% after 5 years of therapy.

Telbivudine

In 2006, telbivudine (L-deoxythymidine (LdT)), a synthetic unsubstituted L-nucleoside analogue of thymidine, was approved for the treatment of CHB at a dose of 600 mg/day. L-Nucleosides differ in their stoichiometric configuration compared with natural nucleosides, with their sugars and base moieties arranged in the L configuration rather than the D configuration. Two years of therapy with telbivudine was superior to lamivudine in both HBeAg-positive and HBeAg-negative patients for achieving therapeutic response, the primary efficacy end point. The intent-to-treat analysis at 2 years showed that telbivudine treatment was also superior to lamivudine for all direct measures of antiviral efficacy, including reduction in serum HBV DNA level from baseline, rates of PCR negativity, and viral resistance, consistent with the 1-year results from this study. [21] The rate of emergence of LdT resistant HBV is lower than that of lamivudine but is still substantial. In a phase III trial that compared LdT with lamivudine, genotypic resistance occurred in 4.4% and 21% of HBeAg-positive patients after 1 and 2 years of treatment, respectively, and 2.7% and 8.6% in HBeAg-negative patients, respectively. The M204I signature mutation was the primary basis for telbivudine resistance, with secondary mutations detected at the L80, L180, and other codons. [22]

Tenofovir

Tenofovir disoproxil fumarate (TDF) is an acyclic nucleotide analogue that is structurally related to adefovir dipivoxil. It has a similar mechanism of action, potency, and antiviral resistance profile as adefovir. The prescribed dose is ≧30-fold greater than adefovir, which may account for its higher potency in vivo. Like adefovir, tenofovir is also effective against the primary lamivudine-resistant virus. ADV-resistant HBV variants may further become selected during TDF treatment, however they cause only a mild decrease in TDF susceptibility. [23] rtA194T polymerase mutation is associated with partial tenofovir drug resistance and negatively impacts replication competence of HBV constructs. Viral replication, however, can be restored to wild type levels, if these polymerase mutations occur together with precore or basic core promoter substitutions as found in HBeAg-negative hepatitis B. Patients with HBeAg-negative chronic HBV infection may therefore be at particular risk when developing drug resistance to tenofovir. Telbivudine or entecavir should be considered as effective alternative treatment options for these patients. [24]

Cross-Resistance among NAs

Cross-resistance is defined as resistance to drug(s) to which a virus has never been exposed. From a cross-resistance perspective, the 5 approved NAs have been placed, based on structural characteristics, into 3 groups: L-nucleosides (LMV and LdT), alkyl phosphonates (ADV and TDF), and D-cyclopentane group (ETV). Resistance and cross-resistance tend to be structure-specific. [25] Therefore, patients who have lamivudine-resistant mutations will not respond to other L-nucleosides, such as telbivudine, whereas the acyclic nucleotides, such as adefovir and tenofovir, are active against lamivudine-resistant mutations. Entecavir of the cyclopentane NA group, although less active against lamivudine-resistant mutants than the wild-type virus in vitro, is still effective in lamivudine-resistant HBV patients at a higher dose. However, entecavir resistance occurs most commonly in the background of lamivudine resistance. In some instances, NAs of the same class may exhibit different resistance profiles. Tenofovir is active against some of the adefovir-resistant mutants (rtA181T/V and rtN236T/D). This lack of cross-resistance can be explained by the relative lower resistance level of mutations induced by adefovir and the higher clinical potency of tenofovir than adefovir. Occasionally, lamivudine treatment can be associated with emergence of rtA181T and rtQ215S mutations, which confer cross-resistance to adefovir and tenofovir. This cross-resistance may explain the more rapid emergence of adefovir resistance in previously lamivudine-treated patients. Therefore, a different class of NAs should be chosen to treat patients with resistance to one class, and combination therapy should be based on selection of NAs with different resistance profiles. [26]

SIDE EFFECT OF LONG-TERM ORAL ANTIVIRAL THERAPY FOR HEPATITIS B

The generally favorable side-effect profiles of the five approved agents during registration trials coupled with the low rates of antiviral drug resistance with the newer agents make them attractive for use beyond 1 year. However, infrequent but serious adverse events such as myopathy, neuropathy, and pancreatitis as well as reversible renal impairment have been reported during postmarketing surveillance. All nucleoside analogs have a "Black Box" warning because of their potential for inhibition of human DNA polymerase gamma involved in mitochondrial DNA (mtDNA) replication. Some of the five approved oral antiviral agents for treatment of HBV have a low level of activity against the human mitochondrial DNA polymerase gamma and can lead to impaired mitochondrial replication with mitochondrial loss or dysfunction. The main function of mitochondria in cells is to oxidize fatty acids and pyruvate to adenosine triphosphate (ATP). Under physiological circumstances, there is a tight coupling of oxidation and phosphorylation that is governed by mitochondrial enzymes such as cytochrome oxidases. Mitochondria also generate reactive oxygen species that can lead to oxidative stress if there is an imbalance between production and intracellular antioxidant defenses. Inhibition of mtDNA polymerase gamma by nucleoside analogs can lead to depletion of intracellular mtDNA with resultant impairment of oxidative phosphorylation and cellular damage.

Lamivudine was the first oral nucleoside analogue approved for the treatment of CHB. It is used at a dose of 100 mg/day. Lamivudine has the high rate of drug resistance mutations, with 76% having genotypic resistance after 8 years of treatment. Therefore lamivudine is considered a second-line therapy for treatment-naive patients. Hepatitis flares occurred in 10% of the lamivudine-treated patients in year 1 and in 18%–21% in years 2–5. A temporal association between hepatitis flares and lamivudine-resistant mutations increased from 43% in year 1 to >80% in year 3. Ten hepatic decompensation events occurred in 8(<1%) lamivudine-treated patients. [27]

Adefovir is an oral nucleotide analogue approved for the treatment of chronic hepatitis B at a dose of 10 mg/day in 2002. Previous studies indicated that nephrotoxicity was the most important dose-limiting toxicity of adefovir dipivoxil therapy observed when studied in patients with HIV type 1 (HIV-1) infection at doses of 60 and 120 mg/d. Nephrotoxicity, as defined by an increase $\geqq 0.5$ mg/dL from baseline in serum creatinine or a serum phosphorus value of<1.5 mg/dL on two consecutive occasions, was not observed in patients treated with ADV at 10 mg for a median follow-up period of approximately 64 weeks. Mild nephrotoxicity was demonstrated with the dose of 30 mg daily. [28] However, in a cohort of 125 patients with chronic HBeAg-negative hepatitis B treated with adefovir for 5 years, the frequency of serum creatinine elevations that were 0.5 mg/dL above baseline was 3%. 8% of the 65 patients who were HBeAg-positive and receiving adefovir for 5 years had reversible creatinine elevations, 5% had albuminuria, and 3% developed hypophosphatemia. Other severe adverse events occurring in two patients (3%) each were abdominal pain, asthenia, fever, lack of drug effect, liver damage, nausea, aspartate aminotransferase (AST) increased, and pharyngitis.

The side effects of adefovir dipivoxil on mitochondria and renal tubular transporter may be involved in the nephrotoxicity. [29] The pathologic changes resulted from nephrotoxicity were mainly extensive edema of epithelium in proximal convoluted tubule, cellular necrosis and vacuolization.

Entecavir is an oral nucleoside analogue approved in 2005 at a dose of 0.5 mg/day for treatment-naive patients and 1.0 mg/day for patients with lamivudine-resistant HBV. Safety analysis of more than 1000 patients in the long-term rollover study demonstrated that entecavir is well tolerated, with 5% adverse events potentially associated with the nucleoside analog. There were 12 patients with confirmed creatinine increases 0.5 mg/dL. Of these, 10 had at least one known renal risk factor. No cases of lactic acidosis were observed.[30] There has been no evidence of mitochondrial or other serious adverse events in patients treated with entecavir for up to 5 years. In cell culture, entecavir demonstrated no evidence of mitochondrial toxicity and did not increase the likelihood of mitochondrial toxicity when combined with lamivudine, adefovir or tenofovir. [31] In the study of 1,633 treated patients, 245 had advanced liver fibrosis/cirrhosis (120 entecavir and 125 lamivudine). No entecavir-treated patients with advanced liver fibrosis/cirrhosis discontinued the therapy due to adverse events. [32]

Telbivudine is a potent oral nucleoside analogue recently approved for treatment of chronic hepatitis B in 2006 at a dose of 600 mg/day. During 4-year telbivudine treatment in patients with chronic hepatitis B, myopathy/myositis, which is an identified risk of telbivudine treatment, was reported in only four patients. Myalgia was the most frequent muscular adverse event (4.6%). Elevated serum creatinine kinase (grade 3 or 4) was reported in 15.9% of patients over 4 years, but remained transient and improved without study drug discontinuation. [33] A novel finding in the present study which is of significant clinical importance is that telbivudine-treated patients show a steady increase in estimated glomerular filtration rate (eGFR) over the 2-year feeder study, remaining at an elevated level thereafter, especially in patients with mild abnormal eGFR at baseline. Improved renal function was maintained for 4-6 years. Increased eGFR with telbivudine treatment was also observed in patients at increased risk for renal impairment. [34] The improvement of glomerular filtration is particularly important in patients with high renal risk such as mildly impaired renal function, decompensated cirrhosis, proteinuria, glomerulonephritis, uncontrolled diabetes or concomitant nephrotoxic drugs.

CONCLUSION

Effective antiviral therapy can reduce the risk of liver cirrhosis and HCC, thus critical for patient care. Drug resistance limits the long-term effectiveness and is a serious problem we face. Factors associated with resistance include dynamics of viral production and clearance, fidelity and efficiency of the viral polymerase, patient compliance, genetic factors that relate to drug metabolism, and features of the antiviral agent itself, including structure and cross-resistance profile. [26] Hepatitis B patients receiving antiviral treatment should be monitored regularly for initial response, adverse events, and breakthroughs. Patients with virological breakthrough should be checked for medication compliance. Whenever possible, testing for resistance mutation should be performed. This is particularly important in patients who have been exposed to more than one nucleos(t)ide analogue therapy. Management of patients with antiviral-resistant HBV depends on prior treatment received, pattern of mutations found, and knowledge of any cross-resistance with other antiviral agents. Recent studies demonstrated that rescue therapy is more effective if initiated early at the time of virological breakthrough, and, in most instances, rescue therapy should be added to the original therapy. [35]

REFERENCES

[1]   Lee WM. Hepatitis B virus infection. *N Engl J Med* 1997;337:1733-45.

[2]   Chen CJ, Yang HI, Su J, et al. Risk of hepatocellular carcinoma across a biological gradient of serum hepatitis B virus DNA level. *JAMA* 2006;295:65-73.

[3]   Fung J, Lai CL, Yuen MF. Hepatitis B and C virus-related carcinogenesis. *Clin Microbiol Infect* 2009;15:964-70.

[4]   Seeger C, Mason WS. Hepatitis B virus biology. *Microbiol Mol Biol Rev* 2000;64:51-68.

[5]   Liang TJ. Hepatitis B: the virus and disease. *Hepatology* 2009;49:S13-21.

[6]   Maillard P, Pillot J. Polymerase chain reaction to monitor repair of the HBV genome, the first step in viral replication. *Res Virol* 1996;147:5-16.

[7]   Fung J, Lai CL, Seto WK, Yuen MF. Nucleoside/nucleotide analogues in the treatment of chronic hepatitis B. *J Antimicrob Chemother* 2011;66:2715-25.

[8] Ridruejo E, Adrover R, Silva MO. Virological breakthrough and resistance in patients with chronic hepatitis B receiving nucleos(t)ide analogues in clinical practice. *Hepatology* 2011;54:1104-5, 1105.

[9] Nowak MA, Bonhoeffer S, Hill AM, Boehme R, Thomas HC, McDade H. Viral dynamics in hepatitis B virus infection. *Proc Natl Acad Sci USA* 1996;93:4398-402.

[10] Ghany M, Liang TJ. Drug targets and molecular mechanisms of drug resistance in chronic hepatitis B. *Gastroenterology* 2007;132:1574-85.

[11] Allen MI, Deslauriers M, Andrews CW, et al. Identification and characterization of mutations in hepatitis B virus resistant to lamivudine. Lamivudine Clinical Investigation Group. *Hepatology* 1998;27:1670-7.

[12] Li MW, Hou W, Wo JE, Liu KZ. Character of HBV (hepatitis B virus) polymerase gene rtM204V/I and rtL180M mutation in patients with lamivudine resistance. *J Zhejiang Univ Sci B* 2005;6:664-7.

[13] Pallier C, Castera L, Soulier A, et al. Dynamics of hepatitis B virus resistance to lamivudine. *J Virol* 2006;80:643-53.

[14] Ridruejo E, Adrover R, Silva MO. Virological breakthrough and resistance in patients with chronic hepatitis B receiving nucleos(t)ide analogues in clinical practice. *Hepatology* 2011;54:1104-5, 1105.

[15] Hadziyannis SJ, Tassopoulos NC, Heathcote EJ, et al. Long-term therapy with adefovir dipivoxil for HBeAg-negative chronic hepatitis B for up to 5 years. *Gastroenterology* 2006;131:1743-51.

[16] Perrillo R, Hann HW, Mutimer D, et al. Adefovir dipivoxil added to ongoing lamivudine in chronic hepatitis B with YMDD mutant hepatitis B virus. *Gastroenterology* 2004;126:81-90.

[17] Angus P, Vaughan R, Xiong S, et al. Resistance to adefovir dipivoxil therapy associated with the selection of a novel mutation in the HBV polymerase. *Gastroenterology* 2003;125:292-7.

[18] Schildgen O, Sirma H, Funk A, et al. Variant of hepatitis B virus with primary resistance to adefovir. *N Engl J Med* 2006;354:1807-12.

[19] Ghany M, Liang TJ. Drug targets and molecular mechanisms of drug resistance in chronic hepatitis B. Gastroenterology 2007;132:1574-85.

[20] Tenney DJ, Levine SM, Rose RE, et al. Clinical emergence of entecavir-resistant hepatitis B virus requires additional substitutions in virus already resistant to Lamivudine. *Antimicrob Agents Chemother* 2004;48:3498-507.

[21] Liaw YF, Gane E, Leung N, et al. 2-Year GLOBE trial results: telbivudine Is superior to lamivudine in patients with chronic hepatitis B. *Gastroenterology* 2009;136:486-95.

[22] Zhao S, Tang L, Fan X, Chen L, Zhou R, Dai X. Comparison of the efficacy of lamivudine and telbivudine in the treatment of chronic hepatitis B: a systematic review. *Virol J* 2010;7:211.

[23] van Bommel F, Trojan J, Deterding K, et al. Evolution of adefovir-resistant HBV polymerase gene variants after switching to tenofovir disoproxil fumarate monotherapy. *Antivir Ther* 2012;17:1049-58.

[24] Amini-Bavil-Olyaee S, Herbers U, Sheldon J, Luedde T, Trautwein C, Tacke F. The rtA194T polymerase mutation impacts viral replication and susceptibility to tenofovir in hepatitis B e antigen-positive and hepatitis B e antigen-negative hepatitis B virus strains. *Hepatology* 2009;49:1158-65.

[25] AegeanSoftware. NoteExpress. In. 2.0 ed; 2005:NoteExpress is a perfect assistant and information manager for researchers, scholars, students, and librarians. NoteExpress is designed to help you organize research notes and bibliographic references, generate bibliographies automatically, search and capture bibliographic data from Internet with efficiency and ease. NoteExpress is well integrated with Microsoft Word. It can format bibliographies in many popular styles.

[26] Ceylan B, Yardimci C, Fincanci M, et al. Comparison of tenofovir and entecavir in patients with chronic HBV infection. *Eur Rev Med Pharmacol Sci* 2013;17:2467-73.

[27] Lok AS, Lai CL, Leung N, et al. Long-term safety of lamivudine treatment in patients with chronic hepatitis B. *Gastroenterology* 2003;125:1714-22.

[28] Izzedine H, Hulot JS, Launay-Vacher V, et al. Renal safety of adefovir dipivoxil in patients with chronic hepatitis B: two double-blind, randomized, placebo-controlled studies. *Kidney Int* 2004;66:1153-8.

[29] Izzedine H, Launay-Vacher V, Deray G. Antiviral drug-induced nephrotoxicity. *Am J Kidney Dis* 2005;45:804-17.

[30] Tang H, Griffin J, Innaimo S, Lehman-McKeeman L, Llamoso C. The Discovery and Development of a Potent Antiviral Drug, Entecavir, for the Treatment of Chronic Hepatitis B. *Journal of Clinical and Translational Hepatology* 2013;1:51-8.

[31] Mazzucco CE, Hamatake RK, Colonno RJ, Tenney DJ. Entecavir for treatment of hepatitis B virus displays no in vitro mitochondrial toxicity or DNA polymerase gamma inhibition. *Antimicrob Agents Chemother* 2008;52:598-605.

[32] Schiff E, Simsek H, Lee WM, et al. Efficacy and safety of entecavir in patients with chronic hepatitis B and advanced hepatic fibrosis or cirrhosis. *The American journal of gastroenterology* 2008;103:2776-83.

[33] Wang Y, Thongsawat S, Gane EJ, et al. Efficacy and safety of continuous 4-year telbivudine treatment in patients with chronic hepatitis B. *J Viral Hepat* 2013;20:e37-46.

[34] Gane EJ, Deray G, Liaw YF, et al. Telbivudine improves renal function in patients with chronic hepatitis B. *Gastroenterology* 2014;146:138-46.

[35] Lok AS. Navigating the maze of hepatitis B treatments. *Gastroenterology* 2007;132:1586-94.

In: Cytochrome P450 Enzymes
Editor: Jian Wu

ISBN: 978-1-61942-209-4
© 2014 Nova Science Publishers, Inc.

Chapter 6

METABOLISM, INTERACTION AND USE GUIDELINES OF ANTI-HBV AND ANTI-HCV MEDICATIONS

Jinsheng Guo[1,2,3] *and Jian Wu*[1,3]*

[1]Key Laboratory of Medical Molecular Virology,
Fudan University Shanghai Medical College, Shanghai, China
[2]Department of Gastroenterology
[3]Shanghai Institute of Liver Disease,
Fudan University Affiliated Zhongshan Hospital,
Shanghai, China

ABSTRACT

New nucleotide derivatives of anti-HBV medications have become the first-line treatment for naïve patients with high viral titers, lamivudine-resistant, abnormal or nearly normal alanine aminotransferase (ALT), yielded a convincing decrease of the viral load, improved HBeAg sero-conversion, and reduced drug resistance development. The treatment significantly minimizes the possibility of progression to end-stage liver disease, decompensated complications and the incidence of hepatocellular carcinoma (HCC). Monotherapy of anti-HBV nucleotides is often well-

---

* Corresponding address: Jian Wu, MD, PhD, Key Laboratory of Molecular Virology, Fudan University Shanghai Medical College, 138 YiXue Yuan Road, Shanghai 200032, China. Email: jian.wu@fudan.edu.cn.

tolerated, and a combination of tenofovir or adefoviror with telbivudine or entecavir is suggested in rescue therapy for those who developed resistance to initial therapy. Thus, it is essential to understand the metabolism and drug interaction when more than one medication is given, especially when patients have existing conditions, such as organ transplantation, HIV co-infection, diabetes, hypertension or renal insufficiency. PEG-IFN plus ribavirin has been the standardized therapy for naïve and genome type I HCV infection with a moderate sustained viral response (SVR). Since adding a protease inhibitor (either boceprevir or telaprevir), the SVR has been elevated approximately 30% in naïve patients, even in those who failed to respond to the standardized regimen in their initial trial. However, both boceprevir and telaprevir are metabolized by cytochrome P450 (CYP) 3A, and the triple therapy regimen may enhance the severity of adverse effects caused by PEG-IFN and ribavirin, such as anemia, thyroid dysfunction, etc. due to drug interaction. When patients receiving the triple regimen have conditions mentioned above, the safe use of anti-HCV and other medications remains a critical issue in their clinical management. Understanding the principle of pharmacology, metabolism and interaction of all prescribed medications, as well as possible factors in potentiating adverse effects or toxicity would greatly improve the successful rate of selected regimens, limit adverse effects in a manageable scale, reduce the drop-out rate, and maintain patient compliance. This chapter will discuss the treatment regimens for HBV and HCV infection, recommended by internationally well-credited professional organizations (such as AASLD, EASL and APASL), provide the use guidelines for preventing possible drug interactions and severe adverse effects.

Keywords: Anti-hepatitis medication, Direct-acting antiviral agents (DAAs), Drug metabolism, Drug interaction, Cytochrome P450

ABBREVIATIONS

AUC, area under the curve
HBV, hepatitis B virus
CYP, cytochrome P450 enzyme
DAA, directing acting anti-viral agents
HBeAg, HBV e antigen
HCC, hepatocellular carcinoma
HCV, hepatitis C virus
NDs, nucleotide derivatives

OLT, orthotopic liver transplantation
PEG-IFN, Pegylated interferon
SVR, sustained viral response

INTRODUCTION

HBV and HCV are the most epidemic viral infection in the world, and both viruses cause chronic hepatitis, which progresses to end-stage liver disease (ESLD) or hepatocellular carcinoma (HCC), the third deadly malignancy. The availability of HBV vaccination brings a hope to eradicate this viral infection in the next few generations in the well-vaccinated areas. However, no HCV vaccination is available due to its frequent genomic mutation. There are approximately 600 million cases of chronic infection with HBV and HCV, and antiviral therapy is an only effective approach to suppress viral load, ameliorate the necro-inflammatory process and complications, reverse fibrosis, delay or block the progression to the ESLD, and lower HCC risk.

Nucleotide derivatives (NDs) for anti-HBV therapy are available. New NDs are developing in the pipelines, and become the first line of treatment for naïve patients. These NDs include tenofovir, adefovir and entecavir. Combination therapy with new NDs (telbivudine) is used to treat those who have failed to respond to an initial regimen or develop resistance to an old generation of NDs, such as lamivudine. NDs are less metabolized by the liver, but they are eliminated through kidneys. Thus, when a combination regimen is prescribed, cautious use of these new NDs is essential.

The availability of protease inhibitors such as boceprevir and telaprevir is a new era in anti-HCV therapy, and new strategies, such as triple therapy or interferon-free regimen are available or in a stage of clinical trials. How to use these direct acting antiviral agents (DAAs) in patients with variety of genome types or confounding conditions, such as co-HIV infection, diabetes, hypertension or orthotopic liver transplantation (OLT), cardiovascular or neuro-psychiatric conditions, is a clinical challenge to medical care professionals. This chapter will focus on therapeutic guidelines based on well-credited recommendations, and provides the rationales of necessary precautions when various regimens are considered in patients with these conditions.

CURRENT GUIDELINES FOR ANTIVIRAL THERAPY
OF HEPATITIS B AND C

HBV

One third of the world's population has been exposed to hepatitis B virus (HBV), and 350 million individuals have been chronically infected with HBV. HBV vaccination dramatically reduced the infection in younger generation, and there is a hope that HBV will be eradicated eventually in the next few generations with effective vaccination programs. However, the vaccination may be less effective in preventing HBV vertical transmission. Antiviral therapy is the only option to control and prevent progression of the disease in patients with chronic HBV infection [1]. Current therapeutic strategies are developed to suppress HBV DNA replication, attenuate inflammation to prevent progressive fibrosis and development of HCC but do not intend to eradicate the virus [2]. The treatment follows the guidelines from internationally well-credited professional societies, such as the American Association for the Study of Liver Disease (AASLD), the European Association for the Study of the Liver (EASL) and the Asian Pacific Association for the Study of the Liver (APASL) [3-5], etc., and the treatment regimens may vary in terms of selection of antiviral medications, viral load, HBeAg status, and alanine aminotransferase (ALT) levels in initial treatment [2].

New nucleotide derivatives of anti-HBV medications have become the first-line treatment for naïve patients with high viral titers, lamivudine-resistant, yielded a convincing decrease in the viral load, reduced serum ALT levels, improved liver histology, and minimized development of drug resistance. The treatment significantly decreases the possibility of progression to end-stage liver disease, decompensated complications and the incidence of HCC. Monotherapy of anti-HBV nucleotides, such as tenofovir and entecavir, both of which target HBV polymerase, is often well-tolerated, and a combination of tenofovir or adefovir with entecavir or telbivudine is suggested in a rescue therapy regimen for those who developed resistance to an initial therapy [2]. Telbivudine (LdT) plus adefovir dipivoxil combination achieved a better treatment outcome than entecavir (ETV) monotherapy regarding serum HBeAg conversion in chronic HBV infection and resistance development [6]. These nucleotide derivatives are able to suppress the viremia beyond the detectable levels (using serum viral load less than 2000 I.U./ml as a cutting-off value for sustained viral response), and may gain a serum conversion of HBeAg at a various rate, but have no effect on HBsAg serum conversion due to the existence of cccDNA and host genomic integration [7].

All orally administrated anti-HBV medications, including lamivudine, are mainly eliminated by kidneys, with a low metabolic rate (only 5-10% for lamivudine) in the liver. Thus, renal function affects their elimination, and precautions must be taken when they are prescribed to patients with renal insufficiency. Attention needs to be paid to the fact that renal dysfunction and Fanconi's syndrome associated with hypophosphatemia caused by long-term administration of low-dose adefovir have been reported in recent years [6]. A significant increase in serum creatine kinase (CK) levels was reported in Chinese patients with telbivudine monotherapy, and it may be associated with myopathy characterized by myalgia and an increase in serum CK levels [8]. Similar to telbivudine, a new anti-HBV derivative, clevudine (revovir), a potent agent in suppressing viral load during an 8-month of clinical trial, was found to cause myopathy characterized by progressive proximal muscular weakness. The mechanism underlying muscular damage was linked to the depletion of mitochondrial DNA in skeletal muscles [9]. Therefore, muscular damage during long-term nucleotide therapy against HBV needs to be monitored. On the other hand, these nucleotide derivatives have less influence on any sub-types of human common CYP (e.g. CYP1A2, CYP2C9, CYP2D6 or CYP3A4) activity or inhibit any other drugs whose metabolism is mediated by CYP isozymes in the liver, even under a much higher concentration *in vitro* than that can be observed *in vivo* in human being. Therefore, they are the choices for HBV patients even with active hepatitis, and are suitable for combination therapy with medications that are metabolized in the liver.

PEG-interferon (PEG-IFN) has been used in naïve patients with HBV infection, and may achieve an approximately 30% HBeAg conversion rate over 12-month treatment. The recommended candidate patients are those with high ALT (>2-5 times upper limit normal value) levels, low HBV DNA (<2X10E8 IU/ml) and genotype A. These patients are most likely to benefit with the treatment [4, 5]. Interleukin 28B polymorphisms may predict an HBeAg conversion rate and overall response [2]. PEG-IFN is usually avoided in patients with decompensated liver disease of HBV infection. Unfortunately, IFN-based regimen carries a substantial adverse effect burden, and is gradually replaced with NDs in these patients.

HCV

Approximately 150 million people are living with chronic HCV infection worldwide, in which two third of HCV-infected individuals will develop chronic liver disease and many will progress to cirrhosis and HCC. HCV genotype 1 is the most common in the world; whereas this genome type is also the most difficult to treat. Chronic hepatitis C infection can be treated and its complications are prevented with an effective antiviral regimen, typically with PEG-IFN plus ribavirin, which has been the standardized therapy for naïve and genome type I HCV infection with a moderate sustained viral response (SVR). However, not all patients are eligible for, able to access, tolerate, or respond to this regimen. The major side effects, such as fatigue/insomnia, anemia, neutropenia, skin reaction, etc. are associated with IFN therapy, and are the major reasons for discontinuation of the standard regimen at various stages of HCV treatment [10]. In order to improve the successful rate of anti-HCV therapy, new compounds that target each of the viral proteins encoded by the single-stranded HCV RNA genome have been developed and evaluated. These compounds include inhibitors of three structural proteins (core, E1 and E2), the ion channel protein p7, and six nonstructural proteins (NS2, NS3, NS4A, NS4B, NS5A or NS5B). Boceprevir and telaprevir, NS3 protease inhibitors, were the first direct-acting antiviral agents (DAAs) to receive a regulatory approval for the treatment of HCV. Simeprevir and sofosbuvir were approved by FDA during December 2013 [11]. Since adding either boceprevir or telaprevir, the SVR of so-called triple therapy regimen has been elevated approximately 30% in both naïve patients and those who failed to respond to the standardized regimen in an initial trial [12]. However, more adverse effect events were also recorded accordingly in the triple regimen than the standard regimen of PEG-IFN plus ribavirin [13], probably due to the fact that PEG-IFN may increase telaprevir steady-state minimum plasma concentration by about 22% compared to telaprevir monotherapy [14].

Enzyme Induction and Mechanism of Interaction

The hemoprotein cytochrome P450 (CYP) plays a key role in the oxidation-reduction process of hepatic drug metabolism in the liver by serving as the terminal oxidase. Numerous CYP isoforms have been identified in human liver, and are responsible for catalyzing the bulk of hepatic drugs and xenobiotic metabolism. Of them, CYP1A2, 2A6, 2C9, 2D6, 2E1 and 3A4 appear to be the major isozymes, accounting for approximately 12, 4, 20, 4, 6 and 28%, respectively, of the total human liver CYP content. It is noteworthy that CYP 3A4 alone is responsible for more than 60% of the clinically prescribed drugs metabolized by the liver, and is the liver-specific CYP isozyme. Table 1 lists the substrates, inducers and drugs used for screening human liver CYP 3A4 activity. This CYP isoform metabolizes the majority of medications on market. [15]

Table 1. Substrates, inducers, and drugs used for screening of human liver CYP3A4

Substrates	Acetaminophen, methadone
	alfentanil, cocaine, dihydroergotamine, lidocaine, sufentanil, tetrahysrocannabinol
	amiodarone, quinidine, diltiazem, verapamil
	astemizole, terfenadine,
	diazepam, midazolam, triazolam
	cortisol, cyclosporine, tacrolimus, troleandomycin
	dapsone, rapamycin, macrolides, miconazole, sulfamethoxazole
	paclitaxel, tamoxifen,
	dihydropyridines,
	ethinyl estradiol, gestodene, testosterone, progesterone, mifepristone (RU 486),
	indinavir, ritonavir, saquinavir,
	lovastatin,
	spironolactone
Inducers	Barbiturates, phenytoin, carbamazepine, glucocorticoids, macrolide antibiotics, rifampin
Noninvasive Markers	Erythromycin, 6β-hydroxycortisol
Inhibitors	furanocoumarins (active components in grapefruit juice)

On repeated administration, some of chemically dissimilar drug substrates are able to "induce" CYP activity by enhancing the rate of its synthesis or reducing its rate of degradation. The net impact is an acceleration of metabolism, and usually results in a decrease in the pharmacologic action of the inducer and also of co-administered drugs, and may also exacerbate metabolite-mediated toxicity by reactive metabolites. On the other hand, certain drug substrates may inhibit CYP isozyme activity by capturing the cytochrome heme-iron. Different drugs may compete for the same endogenous substrates, such as glutathione, glucuronic acid, and sulfate for their inactivation. The faster-reaching drug may effectively deplete endogenous substrate levels and impair the metabolism of the slower-reacting drugs. If the latter has a steep dose-response curve or a narrow margin of safety, potentiation of its pharmacologic and toxic effects may occur.

Impact of DAA Metabolism via CYP 3A4 on Drug Interaction

With availability of boceprevir or telaprevir and other protease inhibitors, IFN-free regimens are on clinical trials [12], and will be the direction towards reducing adverse effects, achieving better patient compliance and successfulness in gaining the same SVR without PEG-IFN. As shown in Table 2, in addition to boceprevir and telaprevir, other protease inhibitors or RNA-dependent RNA polymerase inhibitor (sofosbuvir) are also available, and they share metabolic pathways similar to boceprevir or telaprevir, except sofosbuvir which is mainly excreted through kidneys. Therefore, precautions need to be taken when a triple therapy regimen is prescribed for HCV patients. Metabolic or excretion routes of newly developed protease inhibitors and other anti-HCV medications are summarized in Table 2 for a better understanding of their action mechanisms, drug interaction and adverse effects [16]. It is suggested that in patients being considered for DAA therapy, a full review of prescription record of over-the-counter medications, recreational drugs, and dietary and herbal supplements that the patient uses both regularly and occasionally should be undertaken to identify any potential drug interaction.

Boceprevir is a substrate of the transmembrane transporter P-glycoprotein (P-gp) and an inhibitor of the organic anion uptake transporter 1B1 (SLCO1B1, also known as organic aniontransporting polypeptide 1B1 or OATP1B1). The transmembrane transporter includes P-gp, multidrug resistance protein 1, solute carrier organic anion transporter family member 1B1, multidrug resistance-associated proteins (MRP), and ATP-binding cassette sub-family G member 2 (ABCG2, also known as breast cancer resistance protein or BCRP). Boceprevir is mainly metabolized by aldoketo reductases 1C3 and 1C2 to form various metabolites, such as SCH783007, SCH783005, SCH783006 and SCH783004; known collectively as SCH629144, and it is only partially metabolized by CYP3A4 [17]. Thus, it is a moderate CYP3A inhibitor and a weak P-gp inhibitor. Boceprevir inhibits the drug transporters OATP1B1 and BCRP32, and is 75% protein-bound. After a 400 mg dose of boceprevir, the area under the curve (AUC) and maximum concentration (Cmax) of the active form of boceprevir (SCH534128) were increased 32% and 28%, respectively, in those with moderate hepatic impairment (Child–Pugh B) and 45% and 62%, respectively, in those with severe hepatic impairment (Child–Pugh C), compared to individuals with normal liver function. Whereas, renal function does not affect boceprevir's elimination (patients with end-stage renal disease only have 10% decrease in AUC) and no dosage adjustment is necessary for patients with these conditions, such as diabetes with renal impairment [18]. Other DAAs [16] which are impacted by liver and renal function are shown in Table 3.

Table 2. Metabolic or excretion routes of currently used anti-HCV medications

	Route of metabolism or excretion	CYP effects
Protease inhibitors:		
ABT450/ritonavir	CYP3A	CYP3A inhibition by ritonavir
Asunaprevir	Not determined	Moderate inhibitor of CYP2D6; weak inducer of CYP3A4
Boceprevir	CYP3A	Moderate CYP3A inhibitor
Danoprevir/ritonavir	CYP3A	CYP3A inhibition by ritonavir
Faldaprevir	CYP3A	Moderate inhibitor of hepatic and intestinal CYP3A; weak inhibitor of CYP2C9; inhibitor of UGT1A1
Simeprevir	CYP3A	Mild inhibitor of CYP1A2 and intestinal CYP3A
Telaprevir	CYP3A	Strong CYP3A inhibitor
NS5A Inhibitors:		
ABT267	Not determined	Not determined
Daclatavir	CYP3A	Not determined
Ledipasvir	Not determined	Not a CYP inhibitor or inducer
Nucleotide RNA-dependent RNA polymerase inhibitors:		
Mericitabine	Renal	Not determined
Sofosbuvir	Renal	Not determined
Non-nucleoside polymerase inhibitors:		
ABT333	CYP2C8, CYP3A4, and CYP2D6, contribute approximately 60%, 30%, and 10% to ABT-333 metabolism, respectively	
BI 207127	Not determined	Not determined

Table 3. Impact of hepatic or renal impairment on drug concentration

Drug or Drug Group	Hepatic impairment	Renal impairment
NS3 viral protease inhibitors		
Boceprevir	After a single 400 mg dose of boceprevir, the area under the curve (AUC) and maximum concentration (Cmax) of the active form of boceprevir were increased 32% and 28%, respectively, in those with moderate hepatic impairment and 45% and 62%, respectively, in those with severe hepatic impairment, relative to individuals with no impairment.	Boceprevir AUC is only 10% lower in patients with end-stage renal disease requiring hemodialysis than in individuals with normal renal function. No dosage adjustment is necessary for patients with renal impairment [19].
Telaprevir	Telaprevir AUC and Cmax are reduced by 46% and 49%, respectively, in those with moderate hepatic impairment. The reduction in telaprevir AUC and Cmax was less for those with Child–Pugh A hepatic impairment: 15% and 10%, respectively, so no dose adjustment is necessary in these patients [20].	A single-dose study of telaprevir in individuals with creatinine clearances <30 ml/min/1.73 m^2 found a 10% higher Cmax and 21% higher AUC than those without renal impairment, thus no dosage adjustment is necessary for those with mild, moderate, or severe renal impairment
Simeprevir	In eight volunteers with Child–Pugh B cirrhosis, simeprevir AUC and Cmax were increased 2.62-fold and 1.76-fold, respectively, compared with eight volunteers without hepatic impairment, but similar to those observed in persons with Child–Pugh A cirrhosis.	
Faldaprevir	Faldaprevir pharmacokinetics is not altered in patients with Child–Pugh A cirrhosis.	

Drug or Drug Group	Hepatic impairment	Renal impairment
	Child–Pugh A cirrhosis	
NS5A inhibitors		
Daclatasvir	Dose adjustments of daclatasvir do not seem to be necessary in the setting of hepatic impairment [21]. Total daclatasvir plasma AUC and Cmax are lower in patients with hepatic impairment than healthy controls.	Dose adjustment for renal impairment might not be needed owing to the mainly hepatic route of elimination of this agent.
RNA-dependent RNA polymerase inhibitors		
Sofosbuvir	Sofosbuvir exposures are doubled in individuals with moderate to severe hepatic impairment. Interestingly, despite the increase in parent drug concentration and no change in metabolite concentrations [11].	Dose adjustments of sofosbuvir are necessary in patients with renal impairment [11].

Fifty-nine to 76% telaprevir is protein-bound after oral administration [20]. Telaprevir AUC and Cmax are reduced by 46% and 49%, respectively, in those with moderate hepatic impairment. Telaprevir is extensively metabolized in the liver, involving hydrolysis, oxidation and reduction, mainly catalyzed by CYP3A, and majority of the metabolites are eliminated through feces. Thus, any drugs which are metabolized by the same CYP isozyme would affect telaprevir's metabolism in the liver. Given the fact that only 1% of its metabolite is eliminated through kidneys, the renal function usually does not have significant impacts on telaprevir elimination. Therefore, this medication can be used in patients with mild and moderate renal dysfunction.

As known, other medications may interfere with anti-viral therapy through drug interactions. Table 4 summaries potential medications that may alter the metabolism and plasma concentration of boceprevir, telaprevir, simeprevir and sofosbuvir when co-administrated.

Table 4. Drug interactions between antiviral agents with other medications

Drug or Drug Group	Recommended Dosage	Properties promoting drug interaction	Clinically documented interactions
PEG-IFN alfa-2a	180 μg once Weekly, S.C.	An inhibitory effect on the CYP1A2-mediated metabolism	Theophylline [22]
Boceprevir	800 mg every 7–9 h with food	A moderate CYP3A inhibitor and a weak P-gp inhibitor. Inhibits the drug transporters OATP1B1 and BCRP32 and is 75% protein bound.	Immuno-suppressants (Cyclosporin A and Tacrolimus) [23] Statins (Simvastatin and lovastatin) [24] Antidepressants, anxiolytics, hypotics, mood stabilizers, antipsychotics (Triazolam, oral midazolam, St. John's Wort, carbamazepine and pimozide) [25]
Telaprevir	750 mg every 7–9 h with a high (≥20 g) fat meal	A strong CYP3A inhibitor and moderate P-gp inhibitor; inhibit several hepatic and renal transporters. It is 59–76% protein bound.	Immunosuppressants (Cyclosporin, Tacrolimus, Sirolimus) [26] Opioids (Morphine or Methadone) [27] Statins (Simvastatin and lovastatin) [24] Warfarin or rifampin Antidepressants, anxiolytics, hypotics, mood stabilizers, antipsychotics (Triazolam, oral midazolam, St. John's Wort, carbamazepine and pimozide) [25]
Simeprevir	150 mg once daily	A mild inhibitor of CYP3A and CYP1A2 as well as an inhibitor of OATP1B1 and MRP2.	Cyclosporin and Tacrolimus [28]
Faldaprevir	maximal dose of 240 mg once daily	Moderately and weakly inhibits CYP3A and CYP2C9; a substrate for P-gp and MRP2	Midazolam and S-warfarin

Drug or Drug Group	Recommended Dosage	Properties promoting drug interaction	Clinically documented interactions
Daclatasvir	60 mg once daily	a substrate for CYP3A and a substrate and inhibitor of P-gp; 99% protein bound	
Sofosbuvir	400 mg once daily	Undergoes phosphorylation by host enzymes to a uridine triphosphate analogue for its antiviral effects, thus impaired phosphorylation or portal-systemic shunting could influence viral responses to this DAA	Cyclosporin A

Co-infection of HBV or HCV with HIV

Co-infection of HBV or HCV with HIV is common in patients visiting hepatology clinics. The co-infection increased the anti-retroviral therapy failure and mortality rate in HIV patients [29]. The manifestations and treatment regimen become much more complicated in these patients than those with single viral infection. The extent of HIV progression, occurrence of its complications and prior therapeutic histology are the major factors determining the choice of anti-viral regimens. At the same time, drug interactions are the main concerns in a multiple medication regimen, such as a combined antiretroviral treatment (cART) regimen. Non-nucleoside reverse transcriptase inhibitors are primarily inducers of CYP3A enzymes, with efavirenz being the most potent inducer of this class. Efavirenz reduces the AUC of boceprevir and telaprevir by ~50% in healthy volunteers [30]. On the other hand, boceprevir does not alter the AUC of tenofovir disoproxil fumarate (by 30% [30]). Raltegravir, an HIV integrase inhibitor, can be safely combined with either boceprevir or telaprevir [30, 31]. Rilpivirine (TMC278) is an approved antiretroviral agent for the treatment of HIV-1 infection in naive adults. Rilpivirine is predominantly metabolized by CYP3A4. Rilpivirine at 25 mg once daily does not have a clinically relevant effect on exposure of co-administered drugs. Co-administration with anti-HCV protease inhibitors, either boceprevir or telaprevir, resulted in an increase in rilpivirine plasma concentration, but it is not considered clinically relevant, and no dose adjustments are required. However, when strong CYP3A4 inducers, such as rifampin or rifabutin; or proton pump inhibitors, such as omeprazole or famotidine, are co-administrated, a decrease in rilpivirine plasma concentration was observed, which may affect the efficacy of ripivirine in inhibiting HIV replication [32]. In summary, telaprevir and boceprevir have many interactions with antiretroviral agents that might preclude safe combination. Simeprevir seems to have a list of contraindications similar to telaprevir and boceprevir. The interactions of daclatasvir and faldaprevir with antiretroviral agents seem to be manageable with DAA dose modification. Sofosbuvir has the most benign interaction profile of the DAAs studied with antiretroviral agents to date [32]. The critical issue to be kept in mind is that each medication needs to be evaluated separately, monitored and adjusted in an individual patient in clinical practice.

OTHER CONDITIONS AFFECTING ANTI-VIRAL THERAPY

Patients with viral B or C hepatitis may accompany with other conditions, such as hypertension, diabetes, neuro-psychiatric disorders or organ transplant recipients, which need a long-term of medical treatment. In these cases, medications for a particular condition may affect absorption, distribution, metabolism and elimination of antiviral medications. Potential drug interactions may occur when two categories of medications are metabolized with the same CYP subtypes, such as CYP3A4, or both are eliminated from kidneys. Special attention should be paid to those that are metabolized by CYP3A (Table 1) when they are co-administered with DAAs [33]. In general, the inducers may speed up the metabolism of DAAs; whereas, the inhibitors may compete for the metabolic pathways by the same subtypes, as a result, decrease the DAA metabolism. Ideally, A micro-tuning strategy of dosing and frequency of each medication administration should be implemented based on the plasma concentration of a specific DAA in the regimen, and must be individualized, which is often difficult to be integrated into daily practice in community medical service centers or remote areas. Back & Else summarized a list of medications which are contraindicated with either boceprevir or telaprevir and this list should be helpful for daily practice [33]. For particular interest of hepatology practice, treatment of HCV infection in liver transplant recipients and pregnant women as well as the use of contraceptives are discussed in the following sections.

Liver Transplant Recipients with HCV Infection

Orthotopic liver transplantation (OLT) is the only treatment for end-stage liver disease (ESLD), including HCV infection as an etiology. All patients who undergo OLT with detectable serum HCV RNA experience graft reinfection. The progression of hepatic fibrosis in re-infected recipients is in an accelerated pace, and 20-30% of patients advance to cirrhosis within 5 years post OLT. The outcome of transplant patients with cirrhosis is poor, and approximately 40% of these cases progress to a decompensation stage within 1 year [34]. Meanwhile, challenges exist in the treatment of pre-existing HCV infection or re-infection of implanted grafts in recipients. In either case, failing liver condition and co-administration of immune-suppressive agents affect the adverse effects, completion of an entire course and efficacy of antiviral therapy [28]. Among the adverse effects, anemia and neutropenia are common, and ribavirin dose often needs to be reduced, and erythropoetin or granulocyte colony-stimulating factor (GC-SF) is used to minimize the hematological adverse effects [35].

In liver transplant recipients with cyclosporin A (CsA), sirolimus (Sir) and tacrolimus (TAC)-based immunosuppressive agents, triple therapy gives rise to significant drug interaction, and monitoring telaprevir plasma concentration appeared to be necessary to individually adjust its dose. In a recent study, increasing adverse effects of triple therapy was observed in a cohort of 9 subjects with HCV genome type I after transplant compared to those without transplant, and 2 had to stop the regimen due to side effects, and telaprevir dose had to be adjusted. However, the efficacy was confirmed, and SVR was achieved with undetectable HCV viremia in 6 out 9 recipients who completed a 48-week triple therapy. Nevertheless, 2 patients suffered viral breakthrough [35].

For these patients, monitoring plasma telaprevir concentration seemed to be beneficial to individually adjust its dose and control drug interaction [26]. TAC and CsA are calcineurin inhibitors, which are the basis of the immunosuppressive regimen in OLT patients. It has been suggested that CsA may have potential antiviral action because cyclophilin A, the target protein of CsA, is involved in the viral replication of HCV. A meta-analysis of seven studies with 40 or more patients in each group has shown that dual therapy with CsA was more effective than TAC in achieving a SVR. Moreover, fewer interactions have been described with CsA than with TAC during co-administration with a protease inhibitor [34].

Pregnancy and Oral Contraceptives

Ribavirin is teratogenic. Thus, prevention of pregnancy is of paramount importance during ribavirin-based HCV treatment. Boceprevir and telaprevir reduce ethinyl oestradiol AUC by 26% and 28%, respectively [36]. Boceprevir and telaprevir reduce norethindrone AUC by 4% and 11%, respectively [36]. With telaprevir, the reduction in oral contraceptive exposures affected serum gonadotropin concentrations, suggesting loss of contraceptive efficacy [36]. However, this phenomenon was not observed for boceprevir. Despite the lack of effect of boceprevir on norethindrone pharmacokinetics, drosperinone AUC is doubled by boceprevir. Thus, this progestin should be avoided due to the potential for hyperkalemia and increased likelihood of progestin-related adverse effects. Ethinyl oestradiol and norethindrone were increased by 12% and 15%, respectively, by simeprevir. Daclatasvir did not alter the concentrations of ethinyl oestradiol or norgestimate [21]. Faldaprevir increases ethinyl oestradiol and levonorgestrel AUC by 40%. Thus, oral contraceptive efficacy might be compromised with boceprevir and telaprevir, but not with simeprevir, faldaprevir or daclatasvir.

CONCLUSION AND PERSPECTIVES

HBV and HCV are currently most epidemic viral infection in the world, and approximately 500 million individuals are infected. The availability of HBV vaccination brings the hope to eradicate this viral infection in the next few generations in well-vaccinated areas. No vaccination against HCV is available due to frequent mutation of its genome. The nucleotide derivatives against HBV could suppress viremia beyond undetectable levels, and lead to various levels of HBeAg conversion; whereas, they have a less effect on HBsAg positivity. Few cases developed resistance during treatment with nucleotide derivatives, and the duration of the treatment is often difficult to determine because when cessation of the anti-viral therapy takes place, the viremia started to rebound. The HCV therapy has entered a new era with the availability of protease inhibitors, boceprevir and telaprevir, both of which have improved sustained viral response and resulted in the undertaking of clinical trials with interferon-free regimens. However, the combination therapy for those who developed resistance or failed to respond to initial regimens often requires two or more medications for HBV, and adverse effects such as renal impairment, myopathy, etc. occur in these patients. The addition of either boceprevir or telaprevir does increase the adverse effect events compared to the standard anti-HCV regimen of PEG-IFN plus ribavirin. Most protease inhibitors, such as boceprevir, telaprevir, simeprevir and faldaprevir, are CYP3A inhibitors, and their metabolism is affected by many agents that share the same route of metabolism in the liver. Thus, the precautions must be implemented when a triple therapy regimen is considered, especially when underlying conditions require multiple medications to be administered simultaneously. The therapeutic strategies become complicated when there is co-infection of HBV or HCV with HIV, and anti-retroviral agent, such as ripivirine, may have drug interaction when CYP3A inducers or inhibitors are used in the treatment regimen. Anti-HCV treatment in liver transplant recipients is challenging depending on recipient's general condition, graft function and co-administration of immunosuppressive agents, such as CsA, TAC or Sir. The severity and frequency of adverse effects are more severe but manageable when a triple therapy is considered in transplant recipients in combination with immunosuppressive agents. Monitoring plasma concentration of a protease inhibitor in the regimen will aid in adjusting its dose and managing adverse effects in an individual fashion. Standard PEG-IFN plus ribavirin is contraindicated in woman with pregnancy. Obviously, both boceprevir and telaprevir affect the efficacy of contraceptives. When hepatologists, transplant surgeons and pharmacologists work together, more practical guidelines can be implemented, and the clinical use of anti-viral therapy should be undertaken in a safe fashion.

REFERENCES

[1]   Ghany M, Liang TJ. Drug targets and molecular mechanisms of drug resistance in chronic hepatitis B. *Gastroenterology* 2007;132:1574-85.

[2]   Tujios SR, Lee WM. Update in the management of chronic hepatitis B. *Curr Opin Gastroenterol* 2013;29:250-6.

[3]   Lok AS, McMahon BJ. Chronic hepatitis B: update 2009. *Hepatology* 2009;50:661-2.

[4]   Liaw YF, Leung N, Kao JH, et al. Asian-Pacific consensus statement on the management of chronic hepatitis B: a 2008 update. *Hepatol Int* 2008;2:263-83.

[5]   European Association For The Study Of The L. EASL clinical practice guidelines: Management of chronic hepatitis B virus infection. *J Hepatol* 2012;57:167-85.

[6] Tanaka M, Suzuki F, Seko Y, et al. Renal dysfunction and hypophosphatemia during long-term lamivudine plus adefovir dipivoxil therapy in patients with chronic hepatitis B. *J Gastroenterol* 2014;49:470-80.

[7] Delaney WEt. Molecular virology of chronic hepatitis B and C: parallels, contrasts and impact on drug development and treatment outcome. *Antiviral Res* 2013;99:34-48.

[8] You H, Jia J. Telbivudine treatment in chronic hepatitis B: experience from China. *J Viral Hepat* 2013;20 Suppl 1:3-8.

[9] Seok JI, Lee DK, Lee CH, et al. Long-term therapy with clevudine for chronic hepatitis B can be associated with myopathy characterized by depletion of mitochondrial DNA. *Hepatology* 2009;49:2080-6.

[10] Seeff LB, Ghany MG. Management of untreated and nonresponder patients with chronic hepatitis C. *Semin Liver Dis* 2010;30:348-60.

[11] de Kanter CT, Drenth JP, Arends JE, et al. Viral hepatitis C therapy: pharmacokinetic and pharmacodynamic considerations. *Clin Pharmacokinet* 2014;53:409-27.

[12] Lange CM, Jacobson IM, Rice CM, et al. Emerging therapies for the treatment of hepatitis C. *EMBO Mol Med* 2014;6:4-15.

[13] Marks KM, Jacobson IM. The first wave: HCV NS3 protease inhibitors telaprevir and boceprevir. *Antivir Ther* 2012;17:1119-31.

[14] Marcellin P, Forns X, Goeser T, et al. Telaprevir is effective given every 8 or 12 hours with ribavirin and peginterferon alfa-2a or -2b to patients with chronic hepatitis C. *Gastroenterology* 2011;140:459-468 e1; quiz e14.

[15] MA C. Drug Biotransformation. In: *Basic and clinical pharmacology. Katzung BG* (editor). 8th ed. mcGraw-Hill 13.

[16] Kiser JJ, Burton JR, Anderson PL, et al. Review and management of drug interactions with boceprevir and telaprevir. *Hepatology* 2012;55:1620-8.

[17] Jackson A, D'Avolio A, Moyle G, et al. Pharmacokinetics of the co-administration of boceprevir and St John's wort to male and female healthy volunteers. *J Antimicrob Chemother* 2014.

[18] Kiser JJ, Burton JR, Jr., Everson GT. Drug-drug interactions during antiviral therapy for chronic hepatitis C. *Nat Rev Gastroenterol Hepatol* 2013;10:596-606.

[19] Treitel M, Marbury T, Preston RA, et al. Single-dose pharmacokinetics of boceprevir in subjects with impaired hepatic or renal function. *Clin Pharmacokinet* 2012;51:619-28.

[20] Garg V, Kauffman RS, Beaumont M, et al. Telaprevir: pharmacokinetics and drug interactions. *Antivir Ther* 2012;17:1211-21.

[21] Bifano M, Sevinsky H, Hwang C, et al. Effect of the coadministration of daclatasvir on the pharmacokinetics of a combined oral contraceptive containing ethinyl estradiol and norgestimate. *Antivir Ther* 2013:DOI: 10.3851/IMP2718.

[22] Brennan BJ, Xu ZX, Grippo JF. Effect of peginterferon alfa-2a (40KD) on cytochrome P450 isoenzyme activity. *Br J Clin Pharmacol* 2013;75:497-506.

[23] McLarnon A. Liver transplantation: Boceprevir increases levels of ciclosporin and tacrolimus. *Nat Rev Gastroenterol Hepatol* 2012;9:366.

[24] Chauvin B, Drouot S, Barrail-Tran A, et al. Drug-drug interactions between HMG-CoA reductase inhibitors (statins) and antiviral protease inhibitors. *Clin Pharmacokinet* 2013;52:815-31.

[25] Sockalingam S, Tseng A, Giguere P, et al. Psychiatric treatment considerations with direct acting antivirals in hepatitis C. *BMC Gastroenterol* 2013;13:86.

[26] Farnik H, Zimmermann T, Herrmann E, et al. Telaprevir drug monitoring during antiviral therapy of hepatitis C graft infection after liver transplantation. *Liver Int* 2014.

[27] Fudin J, Fontenelle DV, Fudin HR, et al. Potential P-glycoprotein pharmacokinetic interaction of telaprevir with morphine or methadone. *J Pain Palliat Care Pharmacother* 2013;27:261-7.

[28] Tischer S, Fontana RJ. Drug-drug interactions with oral anti-HCV agents and idiosyncratic hepatotoxicity in the liver transplant setting. *J Hepatol* 2014;60:872-84.

[29] van Griensven J, Phirum L, Choun K, et al. Hepatitis B and C co-infection among HIV-infected adults while on antiretroviral treatment: long-term survival, CD4 cell count recovery and antiretroviral toxicity in Cambodia. *PLoS One* 2014;9:e88552.

[30] van Heeswijk RP, Beumont M, Kauffman RS, et al. Review of drug interactions with telaprevir and antiretrovirals. *Antivir Ther* 2013;18:553-60.

[31] de Kanter CT, Blonk MI, Colbers AP, et al. Lack of a clinically significant drug-drug interaction in healthy volunteers between the hepatitis C virus protease inhibitor boceprevir and the HIV integrase inhibitor raltegravir. *Clin Infect Dis* 2013;56:300-6.

[32] Crauwels H, van Heeswijk RP, Stevens M, et al. Clinical perspective on drug-drug interactions with the non-nucleoside reverse transcriptase inhibitor rilpivirine. *AIDS Rev* 2013;15:87-101.

[33] Back D, Else L. The importance of drug-drug interactions in the DAA era. *Dig Liver Dis* 2013;45 Suppl 5:S343-8.

[34] Coilly A, Roche B, Duclos-Vallee JC, et al. Management of HCV transplant patients with triple therapy. *Liver Int* 2014;34 Suppl 1:46-52.

[35] Werner CR, Egetemeyr DP, Lauer UM, et al. Feasibility of telaprevir-based triple therapy in liver transplant patients with hepatitis C virus: SVR 24 results. *PLoS One 2013*;8:e80528.

[36] Garg V, van Heeswijk R, Yang Y, et al. The pharmacokinetic interaction between an oral contraceptive containing ethinyl estradiol and norethindrone and the HCV protease inhibitor telaprevir. *J Clin Pharmacol* 2012;52:1574-83.

In: Cytochrome P450 Enzymes
Editor: Jian Wu

ISBN: 978-1-61942-209-4
© 2014 Nova Science Publishers, Inc.

Chapter 7

ROLE OF CYTOCHROME P450 ENZYMES IN PSYCHOTROPIC DRUG METABOLISM AND CLINICAL IMPLICATIONS

*Irina Piatkov** *and Trudi Jones*

University Clinic and Research Centre, Blacktown Hospital,
Western Sydney Local Health District, New South Wales, Australia

ABSTRACT

Cytochrome P450 (CYP) isozymes are a major determinant of the pharmacokinetic behaviour of psychotropic drugs. Understanding of the mechanisms involved in the development of drug toxicity has improved with the expanding knowledge of CYP functional variations. Responses to psychotropic drugs, even to standard medical drug treatment regimens, can vary significantly between individuals. Similar dosages can have divergent results due to polymorphism in the genes that code for the enzymes responsible for the metabolism of drugs. Despite the fact that many factors influence the effects of toxins and drugs, including age, organ function, environmental factors and the nature and severity of disease, genetic polymorphisms are accountable for most of the variability in drug treatment response. Interactions between psychotropic drugs metabolised through the Phase I Cytochrome P450 system and an

* Correspondence to: Irina Piatkov, PhD, Senior Hospital Scientist in Charge, Blacktown Molecular Research Laboratory, University Clinic and Research Centre – Blacktown, PO Box 792, Seven Hills 1730, NSW, Australia. E-mail: irina.piatkov@health.

individual's variation in enzyme activity should be considered in the evaluation of treatment. Prescribing clinicians should be mindful of the role of Cytochrome P450 in psychotropic drugs metabolising pathways and their variations, and knowledge of the genetic polymorphisms involved in these pathways will assist clinical decision-making.

Keywords: Cytochrome P450, psychotropic drugs, drug metabolism

INTRODUCTION

Therapeutical response to psychotropic drugs is highly variable and some patients will not experience any remission, while many experience dangerous adverse reactions. This broad spectrum of pharmacological effects is creating complications with prescription of psychotropic drugs. In addition, the medical practitioner who initiates treatment for psychiatric disorders has an extensive list of recommended drugs and very often prescribe a combination of drugs. However, the majority of psychotropic agents use the same metabolic pathways and can compete for the same metabolising enzymes, consequently creating metabolic disturbances. Drug-drug interactions in psycho-pharmacotherapy are very common because many patients are treated with multiple medications. Multiple drug prescriptions have increased during the last few years [1] and the prevalence of polypharmacy in psychiatry varies between 13-90% [2].

Understanding of drug pathways and the functional role of metabolising enzymes by prescribers will help in drug-drug or drug-food interaction interpretation and prevention of side effects. Some psychotropic drugs are prescribed by general medical practitioners (GPs) and it is important to introduce prescription guidelines based on scientific knowledge for drug combination and safe medication management. The constructive guidance for psychotropic drug prescription remains the subject of new developments in medicine. Nevertheless, the latest progress in science and biotechnology provides insights into drug pharmacokinetic and pharmacodynamic pathways. Nowadays, drug metabolism can be analysed on the basis of genetic variations in metabolising enzymes and these variations can be tested in medical laboratories.

The FDA provides extensive recommendations and guidelines for drug prescription, including for psychotropic drugs. Based on information provided by the FDA in 'Pharmacogenetic Biomarkers in Drug Labeling' (data downloaded on 5 March 2014) the Cytochrome P450 biomarkers represent nearly 36% (55 from 155) and the majority of them are from psychotropic drugs (32 from 55), Table 1.

Table 1. Pharmacogenomic Cytochrome P450 Biomarkers in Drug Labeling

	Drug	Therapeutic Area	HUGO Symbol	Referenced Subgroup	Labeling Sections
1	**Amitriptyline**	**Psychiatry**	CYP2D6	CYP2D6 poor metabolizers	Precautions
2	**Aripiprazole**	**Psychiatry**	CYP2D6	CYP2D6 poor metabolizers	Clinical Pharmacology, Dosage and Administration
3	**Atomoxetine**	**Psychiatry**	CYP2D6	CYP2D6 poor metabolizers	Dosage and Administration, Warnings and Precautions, Drug Interactions, Clinical Pharmacology
4	Carisoprodol	Rheumatology	CYP2C19	CYP2C19 poor metabolizers	Clinical Pharmacology, Special Populations
5	Carvedilol	Cardiology	CYP2D6	CYP2D6 poor metabolizers	Drug Interactions, Clinical Pharmacology
6	Celecoxib	Rheumatology	CYP2C9	CYP2C9 poor metabolizers	Dosage and Administration, Drug Interactions, Use in Specific Populations, Clinical Pharmacology
7	Cevimeline	Dermatology	CYP2D6	CYP2D6 poor metabolizers	Drug Interactions
8	**Citalopram (1)**	**Psychiatry**	CYP2C19	CYP2C19 poor metabolizers	Drug Interactions, Warnings
9	**Citalopram (2)**	**Psychiatry**	CYP2D6	CYP2D6 poor metabolizers	Drug Interactions
10	**Clobazam**	**Neurology**	CYP2C19	CYP2C19 poor metabolizers	Clinical Pharmacology, Dosage and Administration, Use in Specific Populations
11	**Clomipramine**	**Psychiatry**	CYP2D6	CYP2D6 poor metabolizers	Drug Interactions

Table 1. (Continued)

	Drug	Therapeutic Area	HUGO Symbol	Referenced Subgroup	Labeling Sections
12	Clopidogrel	Cardiology	CYP2C19	CYP2C19 intermediate or poor metabolizers	Boxed Warning, Dosage and Administration, Warnings and Precautions, Drug Interactions, Clinical Pharmacology
13	**Clozapine**	**Psychiatry**	CYP2D6	CYP2D6 poor metabolizers	Drug Interactions, Clinical Pharmacology
14	**Codeine**	**Anesthesiology**	CYP2D6	CYP2D6 poor metabolizers	Warnings and Precautions, Use in Specific Populations, Clinical Pharmacology
15	**Desipramine**	**Psychiatry**	CYP2D6	CYP2D6 poor metabolizers	Drug Interactions
16	Dexlansoprazole (1)	Gastroenterology	CYP2C19	CYP2C19 poor metabolizers	Clinical Pharmacology, Drug Interactions
17	Dexlansoprazole (2)	Gastroenterology	CYP1A2	CYP1A2 genotypes	Clinical Pharmacology
18	**Dextromethorphan and Quinidine**	**Neurology**	CYP2D6	CYP2D6 poor metabolizers	Clinical Pharmacology, Warnings and Precautions, Drug Interactions
19	**Diazepam**	**Psychiatry**	CYP2C19	CYP2C19 poor metabolizers	Drug Interactions, Clinical Pharmacology
20	**Doxepin**	**Psychiatry**	CYP2D6	CYP2D6 poor metabolizers	Precautions
21	**Drospirenone and Ethinyl Estradiol**	**Neurology**	CYP2D6	CYP2D6 poor metabolizers	Clinical Pharmacology, Warnings and Precautions, Drug Interactions
22	Esomeprazole	Gastroenterology	CYP2C19	CYP2C19 poor metabolizers	Drug Interactions, Clinical Pharmacology
23	**Fluoxetine**	**Psychiatry**	CYP2D6	CYP2D6 poor metabolizers	Warnings, Precautions, Clinical Pharmacology
24	Flurbiprofen	Rheumatology	CYP2C9	CYP2C9 poor metabolizers	Clinical Pharmacology, Special Populations
25	**Fluvoxamine**	**Psychiatry**	CYP2D6	CYP2D6 poor metabolizers	Drug Interactions
26	**Galantamine**	**Neurology**	CYP2D6	CYP2D6 poor metabolizers	Special Populations

	Drug	Therapeutic Area	HUGO Symbol	Referenced Subgroup	Labeling Sections
27	**Iloperidone**	**Psychiatry**	CYP2D6	CYP2D6 poor metabolizers	Clinical Pharmacology, Dosage and Administration, Drug Interactions, Specific Populations, Warnings and Precautions
28	**Imipramine**	**Psychiatry**	CYP2D6	CYP2D6 poor metabolizers	Drug Interactions
29	Lansoprazole	Gastroenterology	CYP2C19	CYP2C19 poor metabolizer	Drug Interactions, Clinical Pharmacology
30	Metoprolol	Cardiology	CYP2D6	CYP2D6 poor metabolizers	Precautions, Clinical Pharmacology
31	**Modafinil**	**Psychiatry**	CYP2D6	CYP2D6 poor metabolizers	Drug Interactions
32	**Nefazodone**	**Psychiatry**	CYP2D6	CYP2D6 poor metabolizers	Drug Interactions
33	**Nortriptyline**	**Psychiatry**	CYP2D6	CYP2D6 poor metabolizers	Drug Interactions
34	Omeprazole	Gastroenterology	CYP2C19	CYP2C19 poor metabolizers	Dosage and Administration, Warnings and Precautions, Drug Interactions
35	Pantoprazole	Gastroenterology	CYP2C19	CYP2C19 poor metabolizers	Clinical Pharmacology, Drug Interactions, Special Populations
36	**Paroxetine**	**Psychiatry**	CYP2D6	CYP2D6 poor metabolizers	Clinical Pharmacology, Drug Interactions
37	**Perphenazine**	**Psychiatry**	CYP2D6	CYP2D6 poor metabolizers	Clinical Pharmacology, Drug Interactions
38	**Pimozide**	**Psychiatry**	CYP2D6	CYP2D6 poor metabolizers	Warnings, Precautions, Contraindications, Dosage and Administration
39	Prasugrel	Cardiology	CYP2C19	CYP2C19 poor metabolizers	Use in Specific Populations, Clinical Pharmacology, Clinical Studies
40	Propafenone	Cardiology	CYP2D6	CYP2D6 poor metabolizers	Clinical Pharmacology
41	Propranolol	Cardiology	CYP2D6	CYP2D6 poor metabolizers	Precautions, Drug Interactions, Clinical Pharmacology

Table 1. (Continued)

	Drug	Therapeutic Area	HUGO Symbol	Referenced Subgroup	Labeling Sections
42	**Protriptyline**	**Psychiatry**	CYP2D6	CYP2D6 poor metabolizers	Precautions
43	Quinidine	Cardiology	CYP2D6	CYP2D6 poor metabolizers	Precautions
44	Rabeprazole	Gastroenterology	CYP2C19	CYP2C19 poor metabolizers	Drug Interactions, Clinical Pharmacology
45	**Risperidone**	**Psychiatry**	CYP2D6	CYP2D6 poor metabolizers	Drug Interactions, Clinical Pharmacology
46	Terbinafine	Infectious Diseases	CYP2D6	CYP2D6 poor metabolizers	Drug Interactions
47	**Tetrabenazine**	**Neurology**	CYP2D6	CYP2D6 poor metabolizers	Dosage and Administration, Warnings, Clinical Pharmacology
48	**Thioridazine**	**Psychiatry**	CYP2D6	CYP2D6 poor metabolizers	Precautions, Warnings, Contraindications
49	Ticagrelor	Cardiology	CYP2C19	CYP2C19 poor metabolizers	Clinical Studies
50	Tolterodine	Urology	CYP2D6	CYP2D6 poor metabolizers	Clinical Pharmacology, Drug Interactions, Warnings and Precautions
51	**Tramadol**	**Analgesic**	CYP2D6	CYP2D6 poor metabolizers	Clinical Pharmacology
52	**Trimipramine**	**Psychiatry**	CYP2D6	CYP2D6 poor metabolizers	Drug Interactions
53	**Venlafaxine**	**Psychiatry**	CYP2D6	CYP2D6 poor metabolizers	Drug Interactions
54	Voriconazole	Infectious Diseases	CYP2C19	CYP219 intermediate or poor metabolizers	Clinical Pharmacology, Drug Interactions
55	Warfarin (1)	Cardiology or Hematology	CYP2C9	CYP2C9 intermediate or poor metabolizers	Dosage and Administration, Drug Interactions, Clinical Pharmacology

(summary from the FDA Table of Pharmacogenomic Biomarkers in Drug Labeling; Psychotropic drugs are highlighted).

Due to adverse drug reactions, which can sometimes even be life-threatening, prescription guidelines and recommendations are critical for drug safety in psychiatry. The FDA labelling for psychotropic drugs, metabolised by Cytochrome P450, includes information about the specific action to be taken and it varies from precautions to warnings or dosage adjustment:

- *Warnings:* Atomoxetine, Citalopram, Codeine, Dextromethorphan and Quinidine, Drospirenone and Ethinyl Estradiol, Fluoxetine, Iloperidone, Pimozide, Tetrabenazine, Thioridazine.
- *Dosage and Administration:* Aripiprazole, Clobazam, Iloperidone, Pimozide, Tetrabenazine.
- *Drug Interaction:* Atomoxetine, Citalopram, Clomipramine, Clozapine, Desipramine, Diazepam, Dextromethorphan and Quinidine, Drospirenone and Ethinyl Estradiol, Fluvoxamine, Iloperidone, Imipramine, Modafinil, Nefazodone, Nortriptyline, Paroxetine, Perphenazine, Risperidone, Trimipramine, Venlafaxine.
- *Use in Specific population:* Clobazam, Codeine, Galantamine, Iloperidone.
- *Contraindication:* Pimozide, Thioridazine.
- *Clinical Pharmacology:* Aripiprazole, Clobazam, Clozapine, Codeine, Dextromethorphan and Quinidine, Diazepam, Drospirenone and Ethinyl Estradiol, Fluoxetine, Paroxetine, Perphenazine, Risperidone, Tetrabenazine, Tramadiol.
- *Precaution:* Amitriptyline, Atomoxetine, Codeine, Doxepin, Dextromethorphan and Quinidine, Drospirenone and Ethinyl Estradiol, Fluoxetine, Iloperidone, Pimozide, Protriptyline, Thioridazine.

CYTOCHROME P450

The Cytochrome P450 (CYP) enzymes are part of the Phase 1 detoxification system that evolved to protect an organism from the breakdown of endogenous (lipids, steroidal hormones) and exogenous (drugs, toxic chemicals) organic substances [3]. CYP is a group of oxidative/dealkylating enzymes which are responsible for the primary metabolism of many drugs, accounting for about 75% of the total number of different metabolic reactions [4, 5]. CYP enzymes are localised in the microsomes of many tissues, including the intestines and liver, and participate in hydroxylation or dealkylation of many commonly prescribed psychotropics such as antidepressants and antipsychotics [6, 7]. In principle, the majority of psychotropic drugs are metabolised by the cytochrome P450 (CYP) family of enzymes [8, 9]. CYP-related metabolites of these drugs are highly chemically active molecules, so consequently they are likely to be associated with toxicity or unpredicted adverse reactions [10-12].

Genes for most CYP enzymes are extremely polymorphic, that is, they frequently contain altered DNA which can produce variants of the enzyme (http://www.imm.ki.se/cypalleles) and when this occurs, the activity of the enzyme is affected. The genetic polymorphism in CYP enzymes is a major factor in the individual variability of drug metabolism [13-16].

Because many patients are prescribed more than one drug [2], there is also a risk of drug-drug interactions affecting the apparent activity of CYP enzymes. Several drugs interfere with specific CYP enzymes to reduce or increase their activity. Also, drugs which are metabolised by the same CYP enzyme will compete for the available activity if they are administered concurrently, for example, antipsychotics and tricyclic antidepressants (TCAs) compete for CYP2D6 activity. Metabolism of TCA will be hampered because of the higher affinity of antipsychotics for CYP2D6. In general, any drug metabolised by a specific CYP enzyme has the potential to inhibit the metabolism of other drugs in the same class [17-30].

A complete overview of CYP substrates, inducers and inhibitors can be found at http://www.medicine.iupui.edu/flockhart. Some of the genetic variations that evolved in humans, since migration from East Africa began, have often become concentrated in geographically isolated populations. Genetic analyses of human relatedness have shown that, on average, most genetic variation (86–88%) occurs within a geographic population and only 12–14% differs between geographically distinct populations, such as between Asia and Europe [31-33]. Even these small differences in allele frequencies lead to differences in biological responses to food and the environment, including responses to drugs.

Ethnic differences in drug metabolism might reflect differences in diet and environment that have changed over thousands of years. The drug-metabolising CYP enzymes in humans have evolved due to the interaction between food and environment, as humans adapted by selection for new enzymes that protected them from the plant, animal or environmental toxins they were exposed to. A well-known example is the high prevalence of ultra-metaboliser CYP2D6 alleles in the Ethiopian population, compared with Europeans. The greater variety of alkaloids in the Ethiopian diet possibly played an evolutionary role in the spread of the ultra-metaboliser genotype in this population [34].

Ethnic specificity in estimating the risk of adverse drug reactions can play an important role in diagnosis and treatment. CYP2C19 poor metabolises are prevalent in Asian populations, for example, and have a higher risk of diazepam-induced sedation. It is therefore advised that lower dosages of diazepam be prescribed to Asian patients than those recommended for Caucasians [3].

Ethnicity can be considered when psychotropic drugs are prescribed and knowledge of a patient's ethnic background could make the identification of a predisposition to adverse drug reactions more efficient.

PSYCHOTROPIC DRUGS AND CYTOCHROME P450 ENZYMES

A psychotropic drug is a chemical substance that can alter the central nervous system function and consequently change the cognition, behaviour, perception, mood and consciousness. Many psychoactive substances are abused and it remains both a social and economic burden for health services and society.

Historically these substances were used for recreational purposes to achieve euphoria or alertness, for ritual/spiritual purposes and for their therapeutic effect. Archaeological evidence of the use of plants with psychoactive effects dates back 10,000 years [35] and the chewing of coca leaves dates back over 8000 years ago [36].

Clinically important psychotropic drugs are mostly used as analgesics or psychiatric medications. There are six major classes of psychiatric medications: Antidepressants, Stimulants, Antipsychotics, Mood stabilisers, Anxiolytics and Depressants. The majority of them are metabolised through the Phase 1 detoxification system, which involves Cytochrome P450; examples for some of the main drugs are described in Figures 1 - 4.

The list of psychotropic drugs metabolised by Cytochrome P450 enzymes is extensive. We have summarised data from the Medication Information CIAP database and the result is as follows:

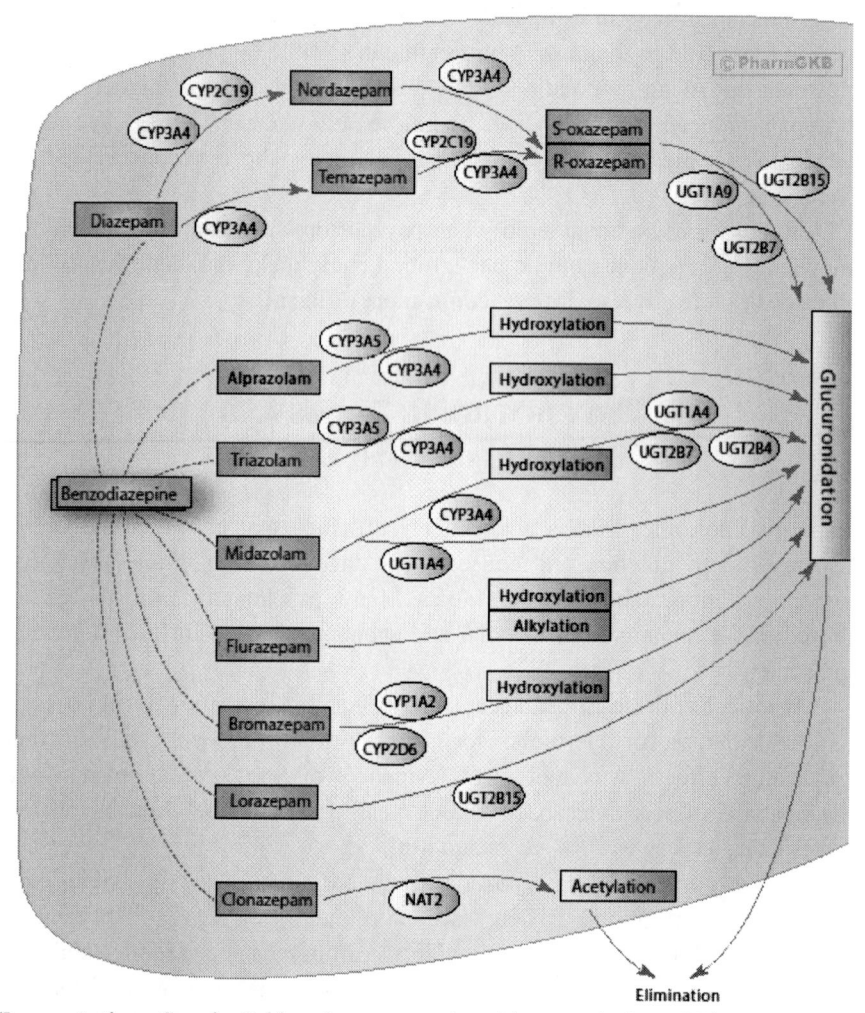

(Image Author: Connie Oshiro; Image reproduced by permission of PharmGKB and
   Stanford University)

There are more than 20 drugs in this class that are used in central nervous system
   disorders, including diazepam. The specific enzymes involved in the
   hydrozylation of BDZs vary, but are primarily CYP3A4 and CYP3A5 and
   CYP2C19, although some studies have found that other CYP enzymes are
   involved [67, 68, 128].

Diagram legend: boxes – drug or metabolite; oval – gene.

Figure 1. Pharmacokinetics of Benzodiazepines (BZDs). Role of Cytochrome P 450
[127].

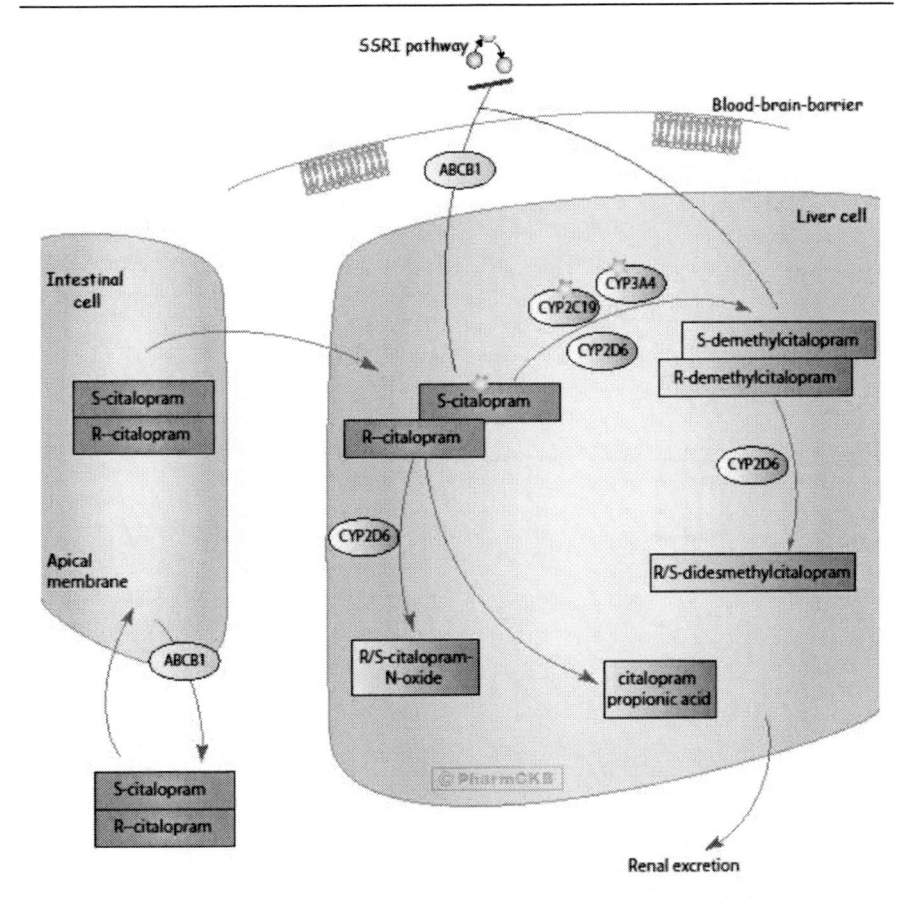

(Image reproduced by permission of PharmGKB and Stanford University)

Both enantimors of citalopram are metabolized by the hepatic cytochrome P450 system. The formation of R/S-demethylcitalopram is catalyzed by the isozymes CYP2C19, CYP3A4, and CYP2D6 [22, 130]. The subsequent N-demethylation to R/S-didesmethylcitalopram is mediated by CYP2D6 [131-133]. Clearance of R/S-citalopram is stereoselective [134-136]. In in vitro studies in human liver microsomes, CYP2C19, CYP3A4, and CYP2D6 all favour the conversion of the biologically active S-enantiomer [132, 135]. Administration of the racemic compounds produces different steady-state concentrations of the R- and S-stereoisomers. Further, N-oxidation and deamination have also been observed and lead to R/S-citalopram N-oxide and citalopram propionic acid metabolites, respectively [137, 138]. The N-oxidation step is also mediated by CYP2D6 [132].

Diagram legend: boxes – drug or metabolite; oval – gene.

Figure 2. Pharmacokinetics of citalopram. Role of Cytochrome P450 [129].

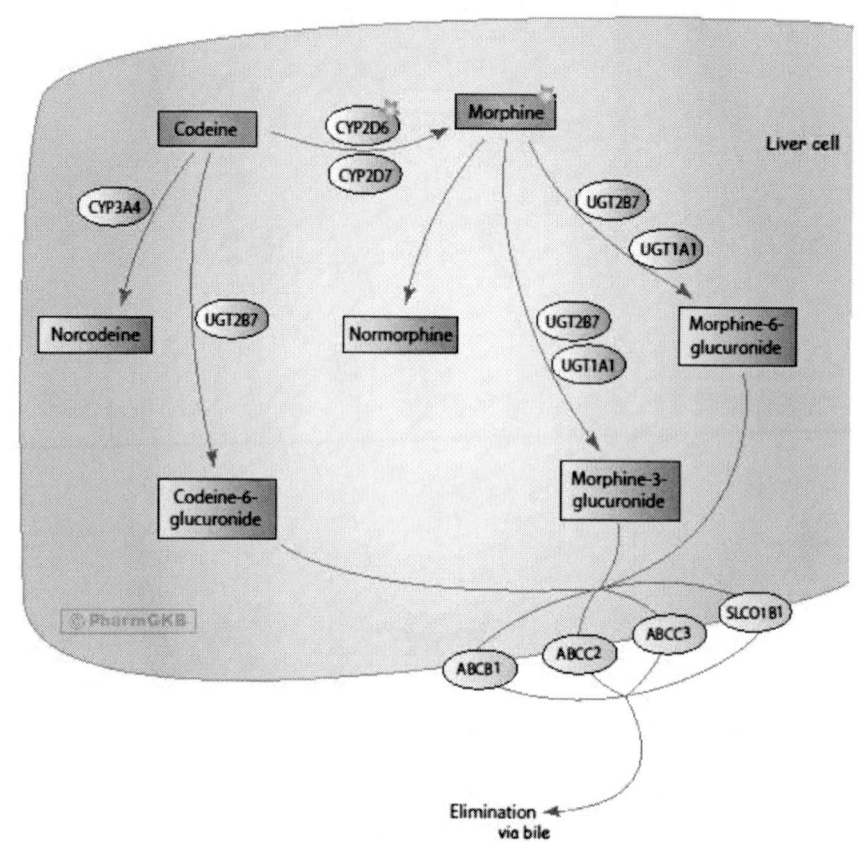

(Image reproduced by permission of PharmGKB and Stanford University)

The principal pathways for metabolism of codeine occur in the liver, although some metabolism occurs in the intestine and brain [139, 140]. Approximately 10-15% of codeine is N-demethylated to norcodeine by CYP3A4. Norcodeine also has a similar affinity to codeine for the mu opioid receptor [141]. Between 0-15% of codeine is O-demethylated to morphine, the most active metabolite, which has 200 fold greater affinity for the mu opioid receptor compared to codeine. This metabolic reaction is performed by CYP2D6 [142-144].

Diagram legend: boxes – drug or metabolite; oval – gene.

Figure 3. Pharmacokinetics of codeine and morphine. Role of Cytochrome P450 [57].

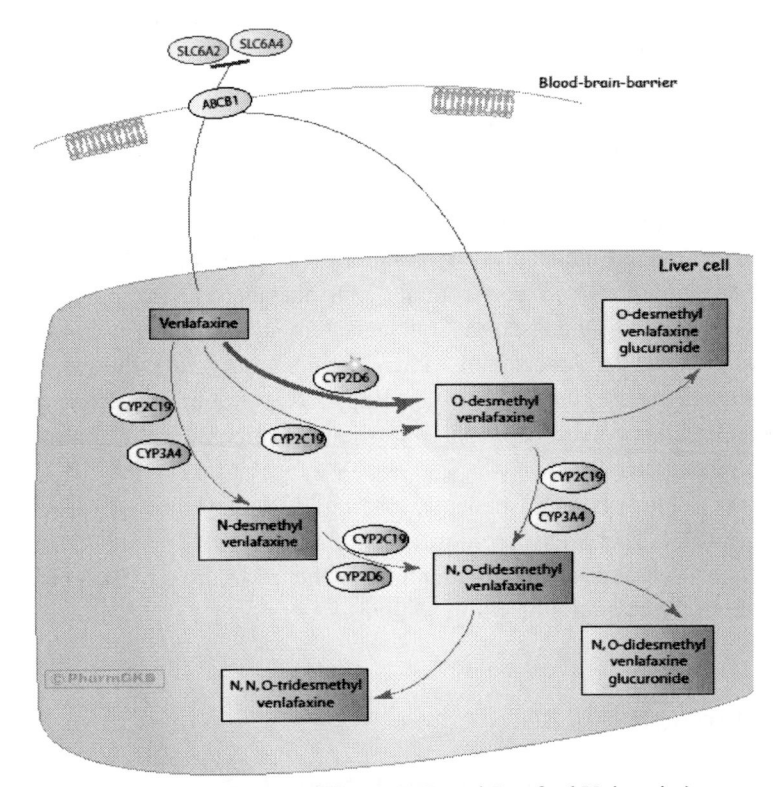

(Image reproduced by permission of PharmGKB and Stanford University)

Demethylation to O-desmethylvenlafaxine (ODV) is the primary route of the first pass metabolism of VEN. Cytochrome P450 2D6 (CYP2D6) is the major enzyme involved in ODV formation [116, 117] A few studies describe a possible stereoselective metabolism of VEN to ODV with either selection towards the (S) isoform [116, 119] or the (R) isoform [118], [146] but the majority of studies regarding VEN pharmacokinetics and antidepressant response in association with CYP2D6 metabolizer phenotype do not distinguish between the enantiomers. Despite the predominate role of CYP2D6, other cytochrome P450 enzymes might be involved in the ODV production to a minor extent [126]. In-vitro experiments implicate the involvement of CYP2C19 in the formation of ODV in human liver microsomes [117]. N-demethylation of VEN to N-desmethylvenlafaxine (NDV) is generally a minor metabolic pathway and catalyzed by CYP3A4 and CYP2C19 [117]. ODV and NDV are further metabolized by CYP2C19, CYP2D6 and/or CYP3A4 to N, O-didesmethylvenlafaxine (NODV), which is a minor metabolite with no known pharmacological effect [115, 123].

Diagram legend: boxes – drug or metabolite; oval – gene.

Figure 4. Pharmacokinetics of Venlafaxine (VEN) is a serotonin-norepinephrine reuptake inhibitor (SNRI). Role of Cytochrome P450 [145].

Cytochrome P450 1A2

Substrates: Agomelatine, Amitriptyline hydrochloride, Asenapine maleate, Caffeine, Clomipramine hydrochloride, Clozapine, Diazepam, Duloxetine, Duloxetine hydrochloride, Fluvoxamine maleate, Imipramine hydrochloride, Levobupivacaine, Levobupivacaine hydrochloride, Lignocaine, Lignocaine hydrochloride, Maprotiline hydrochloride, Mirtazapine, Naproxen sodium, Olanzapine, Olanzapine pamoate monohydrate, Ondansetron, Ondansetron hydrochloride, Palonosetron hydrochloride, Paracetamol, Phenobarbitone, Phenobarbitone sodium, Pimozide, Rasagiline, Riluzole, Ropinirole, Ropinirole hydrochloride, Ropivacaine hydrochloride, Tacrine hydrochloride, Zolmitriptan, Zolpidem tartrate.
Inhibitors: Fluvoxamine maleate, Sertraline, Sertraline hydrochloride, Stiripentol, Tacrine hydrochloride.
Inducers: Carbamazepine, Modafinil, Phenobarbitone, Phenobarbitone sodium, Phenytoin, Phenytoin sodium.

Cytochrome P450 2B6

Substrates: Apomorphine hydrochloride, Bupropion hydrochloride, Methadone hydrochloride.
Inhibitors: Orphenadrine citrate, Orphenadrine hydrochloride.

Cytochrome P450 2C9

Substrates: Amitriptyline hydrochloride, Cannabidiol, Fluoxetine, Fluoxetine hydrochloride, Ibuprofen, Levobupivacaine, Levobupivacaine hydrochloride, Phenytoin, Phenytoin sodium, Tetrahydrocannabinol.
Inhibitors: Amiodarone hydrochloride, Entacapone, Fluoxetine, Fluoxetine hydrochloride, Fluvoxamine maleate.
Inducers: Carbamazepine, Fluoxetine, Fluoxetine hydrochloride, Phenobarbitone, Phenobarbitone sodium, Phenytoin, Phenytoin sodium.

Cytochrome P450 2C19

Substrates: Amitriptyline hydrochloride, Citalopram, Citalopram hydrobromide, Clomipramine hydrochloride, Diazepam, Escitalopram, Escitalopram oxalate, Imipramine hydrochloride, Moclobemide, Phenytoin, Phenytoin sodium, Primidone, Rotigotine.

Inhibitors: Fluoxetine, Fluoxetine hydrochloride, Fluvoxamine maleate, Modafinil, Oxcarbazepine, Stiripentol, Topiramate, Tranylcypromine sulfate.

Inducers: Phenobarbitone, Phenobarbitone sodium, Phenytoin, Phenytoin sodium. Cytochrome P450 2C8

Substrates: Apomorphine hydrochloride.

Cytochrome P450 2D6

Substrates: Amitriptyline hydrochloride, Aripiprazole, Atomoxetine, Atomoxetine hydrochloride, Chlorpromazine hydrochloride, Citalopram, Citalopram hydrobromide, Clomipramine hydrochloride, Clozapine, Codeine, Codeine phosphate, Desipramine hydrochloride, Donepezil, Donepezil hydrochloride, Doxepin, Doxepin hydrochloride, Duloxetine, Duloxetine hydrochloride, Escitalopram, Escitalopram oxalate, Fluoxetine, Fluoxetine hydrochloride, Fluphenazine decanoate, Fluphenazine hydrochloride, Galantamine, Galantamine hydrobromide, Haloperidol, Haloperidol decanoate, Imipramine hydrochloride, Lignocaine, Lignocaine hydrochloride, Maprotiline hydrochloride, Metoclopramide, Metoclopramide hydrochloride, Mianserin hydrochloride, Mirtazapine, Morphine, Morphine hydrochloride, Morphine sulfate, Morphine tartrate, Nortriptyline, Nortriptyline hydrochloride, Olanzapine, Olanzapine pamoate monohydrate, Ondansetron, Ondansetron hydrochloride, Oxycodone, Oxycodone hydrochloride, Oxycodone pectinate, Palonosetron hydrochloride, Paroxetine, Paroxetine hydrochloride, Paroxetine mesilate, Pimozide, Risperidone, Sertindole, Sertraline, Sertraline hydrochloride, Tramadol hydrochloride, Trimipramine, Venlafaxine, Venlafaxine hydrochloride, Zuclopenthixol, Zuclopenthixol acetate, Zuclopenthixol decanoate, Zuclopenthixol dihydrochloride.

Inhibitors: Amiodarone hydrochloride, Asenapine maleate, Bupropion hydrochloride, Citalopram, Citalopram hydrobromide, Codeine, Codeine phosphate, Desipramine hydrochloride, Duloxetine, Duloxetine hydrochloride, Escitalopram, Escitalopram oxalate, Fluoxetine, Fluoxetine hydrochloride, Fluvoxamine maleate, Haloperidol, Haloperidol decanoate, Imipramine hydrochloride, Methadone hydrochloride, Paroxetine, Paroxetine hydrochloride, Paroxetine mesilate, Sertraline, Sertraline hydrochloride.

Inducers: Phenytoin, Phenytoin sodium.

Cytochrome P450 2E1

Substrates: Enflurane, Halothane, Isoflurane, Methoxyflurane, Paracetamol, Sevoflurane,

Cytochrome P450 3A4

Substrates: Alfentanil hydrochloride, Alprazolam, Amiodarone hydrochloride, Amitriptyline hydrochloride, Apomorphine hydrochloride, Aripiprazole, Buprenorphine, Buprenorphine hydrochloride, Buspirone hydrochloride, Caffeine, Cannabidiol, Carbamazepine, Cisapride, Citalopram, Citalopram hydrobromide, Clomipramine hydrochloride, Clonazepam, Clozapine, Codeine, Codeine phosphate, Diazepam, Dihydroergotamine mesylate, Domperidone, Donepezil, Donepezil hydrochloride, Eletriptan hydrobromide, Escitalopram, Escitalopram oxalate, Ethosuximide, Fentanyl, Fentanyl citrate, Galantamine, Galantamine hydrobromide, Haloperidol, Haloperidol decanoate, Levobupivacaine, Levobupivacaine hydrochloride, Lignocaine, Lignocaine hydrochloride, Loratadine, Methadone hydrochloride, Midazolam, Midazolam hydrochloride, Midazolam maleate, Mirtazapine, Nefazodone hydrochloride, Ondansetron, Ondansetron hydrochloride, Oxycodone, Oxycodone hydrochloride, Oxycodone pectinate, Palonosetron hydrochloride, Paracetamol, Phenobarbitone, Phenobarbitone sodium, Phenytoin, Phenytoin sodium, Pimozide, Quetiapine, Quetiapine fumarate, Quinine, Quinine bisulfate, Quinine dihydrochloride, Quinine sulfate, Reboxetine, Reboxetine mesylate, Ropivacaine hydrochloride, Sertindole, Sodium valproate, Sunitinib, Tetrahydrocannabinol, Tiagabine, Tiagabine hydrochloride, Tramadol hydrochloride, Triazolam, Venlafaxine,

Venlafaxine hydrochloride, Ziprasidone, Ziprasidone hydrochloride, Ziprasidone mesilate, Zolpidem tartrate, Zonisamide.

Inhibitors: Amiodarone hydrochloride, Fluoxetine hydrochloride, Fluvoxamine maleate, Nefazodone hydrochloride, Quinine, Quinine bisulfate, Quinine dihydrochloride, Quinine sulfate, Sertraline, Sertraline hydrochloride, Stiripentol.

Inducers: Carbamazepine, Hypericum perforatum, Hypericum perforatum dry extract, Hypericum perforatum dry extract concentrate, Hypericum perforatum dry leaf, Hypericum perforatum flowering herb top dry, Hypericum perforatum flowering herb top extract dry conc., Hypericum perforatum herb dry, Hypericum perforatum oil, Hypericum perforatum stem dry, Modafinil, Oxcarbazepine, Phenobarbitone, Phenobarbitone sodium, Phenytoin, Phenytoin sodium.

MANIFESTATION OF ADVERSE DRUG REACTIONS AND CYTOCHROME P450 VARIATIONS

Psychotropic medications have been associated with a variety of Adverse Drug Reactions (ADRs), including the development of neurotoxicity. As described by Wall and co-authors [37] adverse drug reactions that have been linked to abnormal metabolism of psychotropics and effects associated with abnormal metabolism of drugs are numerous:

a. *Individuals with reduced metabolism*: Extrapyramidal Symptoms, tardive dyskinesia, over sedation, cardiovascular complications, weight gain, neuroleptic malignant syndrome, serotonin syndrome, suicidality.

b. *Individuals with increased metabolism:* opioid toxicity, nausea, paradoxical excitation, treatment nonresponse, suicidality.

In clinical practice, depending on a patient's ability to break down drugs in the body, four different phenotype groups are used to describe variations in drug response based on the genotype of the metabolising enzymes:

- poor metaboliser phenotypes (PMs), with reduced enzyme activity
- intermediate metaboliser phenotypes (IMs)

- extensive metaboliser phenotypes (EMs), with normal enzyme activity
- ultrarapid metaboliser phenotypes (UMs), with increased enzyme activity

Dose adjustment based on pharmacogenetics data is recommended by the FDA (US Food and Drug Administration) and European Committees (European Public Assessment Report and Royal Dutch Pharmacists Association) for several drugs. Recommendations are based on pharmacokinetic parameters, including oral clearance, area under the concentration time curve and concentration at steady state.

A subset (1-10%) of the population has reduced activity of certain Cytochrome P450 drug metabolising enzymes. These individuals may develop higher than expected plasma concentrations of drugs when given usual doses. In addition, certain drugs that are metabolised by the same isozyme (e.g. CYP2D6) may inhibit the activity of this isozyme (refer to the inhibitors list above), and consequently normal metabolisers can be transformed into poor metabolisers during the co-administration of drugs involved in the same pathway. Below we describe some main psychotropic drugs and recommendations based on the role of Cytochrome P450 in their metabolism.

Aripiprazole

An atypical antipsychotic used in the treatment of schizophrenia, bipolar I disorder and manic episodes. It is mainly metabolised by CYP2D6 and CYP3A4 [38-40].

Dose adjustments due to interactions are recommended by the FDA and by the EMA European Public Assessment Report (EPAR) when aripiprazole (ARI) is taken with CYP3A4 or CYP2D6 inhibitors or inducers. The aripiprazole dose should be reduced if ARI is co-administered with CYP3A4 or CYP2D6 inhibitors and the aripiprazole dose should be increased if CYP3A4 inducers are used in combinational treatment [41, 42].

The Royal Dutch Pharmacists Association - Pharmacogenetics Working Group (DPWG) recommends reducing the maximum dose of aripiprazole for patients carrying poor metaboliser alleles of CYP2D6 [43].

Hendsen and co-authors [38] demonstrated that serum concentrations of both aripiprazole and the active sum of ARI and its metabolite in psychiatric patients were significantly affected by CYP2D6 genotype. The observed differences indicate that poor metabolisers typically need 30-40% lower doses to achieve a similar steady-state serum concentration compare with extensive metabolisers. This finding is supported by several publications describing clinical effects of aripiprazole in CYP2D6 PMs [44-46] and intermediate metabolisers [18, 39, 47].

Amitriptyline and Tricyclic Antidepressants (TCAs)

The FDA drug label warns that CYP2D6 poor metabolisers have higher than expected plasma concentrations of tricyclic antidepressants (TCAs) when given usual doses and that certain drugs inhibit the activity of the CYP2D6 isozyme and make normal metabolisers resemble poor metabolisers. "It is desirable to monitor TCA plasma levels whenever a TCA is going to be co-administered with another drug known to be an inhibitor of P450 2D6" [41].

Clinical Pharmacology and Therapeutics Guidelines, by the Clinical Pharmacogenetics Implementation Consortium (CPIC), for amitriptyline or nortriptyline recommends an alternative drug for CYP2D6 or CYP2C19 ultrarapid metabolisers and for CYP2D6 poor metabolisers. A 50% dose reduction for CYP2C19 poor metabolisers and a 25% dose reduction for CYP2D6 intermediate metabolisers are recommended [9].

The Royal Dutch Pharmacists Association - Pharmacogenetics Working Group has evaluated therapeutic dose recommendations for amitriptyline based on CYP2D6 genotypes [43]. An alternative drug, or reducing the initial dose, is recommended for patients carrying intermediate metaboliser alleles and selecting alternative drugs or monitoring amitypityline and nortriptyline plasma concentration for patients carrying the PMs or ultrarapid metaboliser alleles.

Tricyclic antidepressants have comparable pharmacokinetic properties, so it may be reasonable to apply the CPIC Dosing Guideline for amitriptyline and CYP2C19, CYP2D6 to other tricyclics: doxepin, imipramine, protriptyline, trimipramine.

Clozapine

Clozapine is an atypical antipsychotic drug. It is approved for the treatment of severely ill patients with schizophrenia who fail to show an acceptable response to standard antipsychotic drug treatment. Clozapine is a substrate of CYP1A2, CYP3A4 and CYP2D6 [48-52].

The FDA drug label recommends dose adjustments in CYP2D6 poor metabolisers and with concomitant use of strong CYP1A2 inhibitors (e.g. fluvoxamine, ciprofloxacin or enoxacin); moderate or weak CYP1A2 inhibitors (e.g. oral contraceptives or caffeine); CYP2D6 or CYP3A4 inhibitors (e.g. cimetidine, escitalopram, erythromycin, paroxetine, bupropion, fluoxetine, quinidine, duloxetine, terbinafine or sertraline); CYP3A4 inducers (e.g. phenytoin, carbamazepine, St. Johns Wort or rifampin); or CYP1A2 inducers (e.g. tobacco smoke). When co-administering clozapine with other drugs that are metabolised by CYP2D6, the FDA recommends using lower doses of such drugs (i.e. antidepressants, phenothiazines, carbamazepine, propafenone, flecainide and encainide). This recommendation is based on numerous research data [19, 26, 28, 40, 48, 50, 51, 53, 54].

One study has shown, for the first time, a significant in vivo role of CYP2C19 and the P-gp transporter in the pharmacokinetics of clozapine [49], however, the association with genetic polymorphisms needs to be established.

Codeine

Codeine, an opiate analgesic, is used as a cough suppressant, for pain control and as an antidiarrhoeal agent. Conversion to morphine is linked to codeine's analgesic activity. It functions as both positive and negative regulators of synaptic transmission via G-proteins and inhibits the release of nociceptive neurotransmitters such as substance P, GABA, dopamine, acetylcholine and noradrenaline [55]. Opioids also inhibit the release of vasopressin, somatostatin, insulin and glucagon, as they regulate calcium and potassium channels, which results in hyperpolarisation and reduced neuronal excitability [56].

Codeine is metabolised to morphine by CYP2D6 [57]. Several studies demonstrate that CYP2D6 ultrarapid metabolisers may experience adverse drug reactions [58] and mothers who are ultrarapid metabolisers, and breast-feeding, have potentially dangerously high serum morphine levels in their breastfed infants, which can even lead to the infants' death [59]. The FDA added a boxed warning to the label for children who receive codeine following tonsillectomy and/or adenoidectomy who may be rapid metaboliser genotype. In a patient identified as a CYP2D6 ultrarapid metaboliser the FDA recommends an alternative analgesic to avoid the risk of severe toxicity [41, 56, 60].

Patients with a CYP2D6 intermediate metaboliser phenotype should be monitored closely or should be offered an alternative analgesic. As for CYP2D6 poor metabolisers, the use of codeine should be avoided and an alternative analgesic should be used [56, 60, 61].

Quinidine

Quinidine acts on sodium channels on the neuronal cell membrane, limiting the spread of seizure activity and reducing seizure propagation. The antiarrhythmic actions are mediated through effects on sodium channels in Purkinje fibres. Quinidine, a hydantoin anticonvulsant, is used alone or with phenobarbital or other anticonvulsants to manage tonic-clonic seizures, psychomotor seizures, neuropathic pain syndromes, including diabetic neuropathy, digitalis-induced cardiac arrhythmias, and cardiac arrhythmias associated with QT-interval prolongation. Quinidine is a Class I antiarrhythmic medication which inhibits CYP2D6 [62-64].

From the FDA drug label: "Quinidine is not metabolized by cytochrome P450IID6, but therapeutic serum levels of quinidine inhibit the action of cytochrome P450IID6, effectively converting extensive metabolizers into poor metabolizers. Caution must be exercised whenever quinidine is prescribed together with drugs metabolised by cytochrome P450IID6". Therefore, CYP2D6 poor metabolisers may be at risk when quinidine is used in co-administration with other drugs or drug components.

Diazepam

Diazepam (Benzodiazepine) is a drug intended for treatment of breakthrough seizures. Diazepam is metabolised by CYP2C19 and CYP3A4 and inter-individual variation in clearance of the drug is likely attributable to CYP2C19 or CYP3A4 genetic variability [7, 65, 66].

There are more than 20 Benzodiazepines (BDZs) that are used in central nervous system disorders. The specific enzymes involved in the hydrozylation of BDZs vary, but are primarily CYP3A4, CYP3A5 and CYP2C19, although some studies have found that other CYP enzymes are involved [7, 15, 67-69].

Fluoxetine

Fluoxetine is a selective serotonin reuptake inhibitor (SSRI) and is used for the treatment of major depressive disorder (MDD), obsessive-compulsive disorder (OCD), bulimia nervosa and panic disorder [70].

Fluoxetine is metabolised by several cytochrome P450 enzymes with CYP2D6 being a major contributor [16]. At the same time, fluoxetine is an inhibitor of the CYP2D6 enzyme pathway and a potential drug-drug interaction substrate [25]. Co-administration of fluoxetine with other drugs that are metabolised by CYP2D6 can convert a normal CYP2D6 metaboliser to a poor metaboliser.

As recommended by the FDA label [41], "therapy with medications that are predominantly metabolized by the CYP2D6 system and that have a relatively narrow therapeutic index should be initiated at the low end of the dose range if a patient is receiving fluoxetine concurrently or has taken it in the previous 5 weeks. Thus, his/her dosing requirements resemble those of poor metabolizers. If fluoxetine is added to the treatment regimen of a patient already receiving a drug metabolized by CYP2D6, the need for decreased dose of the original medication should be considered".

Fluoxetine's metabolism involves the Cytochrome P450 system, and a combination of drugs which are also metabolised by CYPs may lead to drug interactions, even if fluoxetine has been discontinued for 4-5 weeks [71-76].

Galantamine

Galantamine is a benzazepin and is used to reverse the muscular effects of gallamine triethiodide and tubocurarine, and has been studied as a treatment for Alzheimer's disease and other central nervous system disorders [77].

In vitro studies indicate that cytochrome CYP2D6 and CYP3A4 were the major cytochrome P450 isozymes involved in the metabolism of galantamine and CYP2D6 and CYP3A4 inhibitors increase oral bioavailability of galantamine [78-80]. Pharmacokinetic analysis indicated that there was a 25% decrease in median clearance in poor metabolisers compared to extensive metabolisers [79-82]. However, dosage adjustment is not recommended as the dose of drug is individually titrated to tolerability.

Iloperidone

Iloperidone is an atypical antipsychotic for the treatment of schizophrenia [70].

Elimination of iloperidone is mainly through hepatic metabolism involving two P450 isozymes, namely CYP2D6 and CYP3A4. To avoid increase of iloperidone in plasma, it is recommended that the dosage of iloperidone should be reduced to half when used with known strong inhibitors of CYP2D6 or CYP3A4 or in CYP2D6 poor metabolisers [83-87].

Modafinil

Modafinil is a stimulant drug marketed as a 'wakefulness promoting agent' and is one of the stimulants used in the treatment of narcolepsy. It is metabolised by many cytochrome P450s [88, 89].

As per the FDA label, drugs that are largely eliminated via CYP2C19 metabolism may have prolonged elimination upon co-administration with modafinil and may require dosage reduction and monitoring for toxicity. Modafinil and its metabolite reversibly inhibit CYP2C19 and could potentially modify the metabolism of CYP2C19 substrates [90]. Consequently, co-prescription with Tricyclic Antidepressants is not recommended.

Paroxetine

Paroxetine is used in major depressive disorder (MDD), panic disorder with or without agoraphobia, obsessive-compulsive disorder (OCD), social anxiety disorder (social phobia), generalised anxiety disorder (GAD), post-traumatic stress disorder (PTSD) and premenstrual dysphoric disorder (PMDD). Unlabelled indications include: eating disorders, impulse control disorders, vasomotor symptoms of menopause, obsessive-compulsive disorder (OCD) in children, and mild dementia-associated agitation in nonpsychotic individuals [70].

The Royal Dutch Pharmacists Association - Pharmacogenetics Working Group has evaluated therapeutic dose recommendations for paroxetine based on CYP2D6 genotype and suggests using an alternative drug for ultrarapid metabolisers [91].

Paroxetine inhibits CYP2D6 and consequently inhibits the metabolism of drugs metabolised by CYP2D6. Drug interaction can cause serious complications and, for example, paroxetine should not be co-administrated with thioridazine due to the risk of serious ventricular arrhythmias or co-administration with tamoxifen may lead to reduced efficacy of tamoxifen [24, 92, 93].

Perphenazine

Perphenazine is a piperazinyl phenothiazine and is a highly potent antipsychotic and dopamine receptor antagonist [70].

As recommended by the FDA, poor metabolisers of CYP2D6 will metabolise perphenazine slowly and will experience higher concentrations compared with normal or extensive metabolisers. Co-administration with CYP2D6 inhibitors should be avoided [94, 95].

Pimozide

Pimozide is an antipsychotic drug which blocks dopaminergic receptors on neurons in the central nervous system [70].

Dosage adjustment is suggested by the FDA for individuals with genetic variations resulting in poor CYP2D6 metabolism [41]. Studies show differences in concentrations detected in poor metabolisers than extensive CYP2D6 metabolisers. "The time to achieve steady state pimozide concentrations is expected to be longer (approximately 2 weeks) in poor CYP2D6 metabolisers because of the prolonged half-life". As per the increased pimozide concentrations and longer time to steady state in CYP2D6 poor metabolisers, clinicians should follow the dosing, titration, and genotype testing recommendations for paediatric and adult patients [96, 97].

Risperidone

Risperidone is an atypical antipsychotic drug with high affinity for 5-hydrotryptamine (5-HT) and dopamine D2 receptors. It is used primarily in the management of schizophrenia, inappropriate behaviour in severe dementia and manic episodes associated with bipolar I disorder [70].

Risperidone is mostly metabolised by CYP2D6 [26, 28, 54, 94]. However, despite known variations in the activity of the enzyme in population, data does not suggest that poor and extensive metabolisers have different rates of adverse effects. The FDA suggestion is only made for respiridone drug interaction with CYP2D6 substrates.

The Royal Dutch Pharmacists Association - Pharmacogenetics Working Group has evaluated therapeutic dose recommendations for risperidone based on CYP2D6 genotypes [43]. Based on numerous studies, for patients who are CYP2D6 poor metabolisers, intermediate metabolisers or ultrarapid metabolisers, the Dutch Group recommends selecting an alternative drug or being extra alert to Adverse Drug Events and adjusting dose to clinical response [21, 40, 98-101].

Tetrabenazine

Tetrabenazine (TBZ) is a monoamine-depleting agent and formerly used as an antipsychotic but now used primarily in the treatment of various movement disorders including tardive dyskinesia, dystonias, choreas, dyskinesias and tics [70]. Tetrabenazine is used to treat Huntington's Disease chorea [42]. Its primary metabolites are metabolised mainly by CYP2D6, and people with CYP2D6 poor metaboliser genotypes should be treated with lower doses [102].

The FDA recommends a dosage adjustment according to the CYP2D6 genotype and for prescription combination with strong CYP2D6 inhibitors such as quinidine or antidepressants (e.g. fluoxetine or paroxetine) [102].

Thioridazine

Thioridazine is used to treat schizophrenic patients and in the control of severely disturbed or agitated behaviour. The psychotropic effect of thioridazine is based on the postsynaptic mesolimbic dopaminergic D1 and D2 receptors block in the brain; it blocks alpha-adrenergic effect, depresses the release of hypothalamic and hypophyseal hormones and is believed to depress the reticular activating system thus affecting basal metabolism, body temperature, wakefulness, vasomotor tone and emesis [42, 70].

The main metabolising enzyme is CYP2D6 and patients with reduced CYP2D6 activity should be checked for the CYP2D6 genotype. Due to its potentially fatal effects on heart rhythm, the FDA recommends to use it only if other antipsychotic drugs are not effective or cause intolerable side effects [103-109].

Tramadol

As indicated in the treatment of moderate to severe pain, Tramadol is used to treat postoperative, dental, cancer and acute musculosketetal pain and as an adjuvant to NSAID therapy in patients with osteoarthritis [70].

Tramadol is mostly metabolised by Cytochrome P450 [110-114].

According to the Royal Dutch Pharmacists Association - Pharmacogenetics Working Group tramadol therapeutic dose should be based on CYP2D6 genotypes [43]. Alternative (not oxycodone or codeine) drugs should be selected for patients with CYP2D6 poor metaboliser genotypes and CYP2D6 intermediate metaboliser genotypes. As for CYP2D6 ultrarapid metabolisers, the group recommends 30% decreased dose or use of an alternative to tramadol (not oxycodone or codeine).

Venlafaxine

Venlafaxine is a serotonin-norepinephrine reuptake inhibitor (SNRI). It is used as an antidepressant and prescribed for the treatment of clinical depression and anxiety disorders. It inhibits neuronal serotonin and norepinephrine reuptake and dopamine reuptake [70]. It is mostly metabolised by CYP2D6 and CYP3A4 [115, 116].

Dose adjustment is not recommended for venlafaxine based on CYP2D6 genotypes by the Royal Dutch Pharmacists Association - Pharmacogenetics Working Group for PM and IM genotypes (1) due to insufficient data. However, numerous research reports indicate the use of an alternative drug or dose adjustment to clinical response for UM genotypes (titrating dose to a maximum of 150% of the normal dose) [20, 117-126].

CONCLUSION

Cytochrome P450 enzymes developed in animals to protect the organism from dietary and environmental toxins. The biotransformation of dietary toxins, carcinogens and mutagens are controlled by these enzymes and insufficient activity in these enzymes may lead to the development of specific diseases.

The concentration of many drugs and their metabolites in the organism depends on the activity of Cytochrome P450 enzymes. Additionally, the drug metabolites produced during this transformation are biologically active and can add to the therapeutic effect or toxicity of the parent drug.

A drug itself can increase or decrease the functional activity of CYP enzymes and furthermore, variations in enzyme activity among different patients cause variations in drug response. All these factors should be taken into consideration in clinical practice.

Knowledge of Cytochrome P450 function and its biological variations can help to determine whether patients will respond to treatment, or whether the drug combination will change the drug/metabolites ratio, mediating clinically important effects. Genotype tests are available and can be performed to determine the cytochrome specific activity. Nowadays physicians can have tools to predict a drug-drug interaction and can select the appropriate drug or dosage to avoid side effects.

REFERENCES

[1]   Roh D, Chang JG, Kim CH, Cho HS, An SK, Jung YC. Antipsychotic polypharmacy and high-dose prescription in schizophrenia: a 5-year comparison. *Aust. N. Z. J. Psychiatry* 2014;48(1):52-60.

[2]   Kukreja S, Kalra G, Shah N, Shrivastava A. Polypharmacy in psychiatry: a review. *Mens Sana Monogr.* 2013;11(1):82-99.

[3]   Chang GW, Kam PC. The physiological and pharmacological roles of cytochrome P450 isoenzymes. *Anaesthesia* 1999;54(1):42-50.

[4]   Korzekwa K. Enzyme kinetics of oxidative metabolism: cytochromes p450. *Methods Mol. Biol.* 2014;1113:149-66.

[5]   Hoffmann MF, Preissner SC, Nickel J, Dunkel M, Preissner R, Preissner S. The Transformer database: biotransformation of xenobiotics. *Nucleic Acids Res.* 2014;42(Database issue):D1113-7.

[6]   Attia TZ, Yamashita T, Miyamoto M, et al. Comparison of cytochrome p450 mediated metabolism of three central nervous system acting drugs. *Chem. Pharm. Bull.* (Tokyo) 2012;60(12):1544-9.

[7]   Fukasawa T, Suzuki A, Otani K. Effects of genetic polymorphism of cytochrome P450 enzymes on the pharmacokinetics of benzodiazepines. *J. Clin. Pharm. Ther.* 2007;32(4):333-41.

[8]   Hiemke C, Shams M. Phenotyping and genotyping of drug metabolism to guide pharmacotherapy in psychiatry. *Curr. Drug Deliv.* 2013; 10(1):46-53.

[9]   Hicks JK, Swen JJ, Thorn CF, et al. Clinical Pharmacogenetics Implementation Consortium guideline for CYP2D6 and CYP2C19 genotypes and dosing of tricyclic antidepressants. *Clin. Pharmacol. Ther.* 2013;93(5):402-8.

[10]   Hodgson K, Tansey K, Dernovsek MZ, et al. Genetic differences in cytochrome P450 enzymes and antidepressant treatment response. *J. Psychopharmacol.* 2014;28(2):133-41.

[11]   Turner RM. From the lab to the prescription pad: genetics, CYP450 analysis, and medication response. J. Child Adolesc. Psychiatr. Nurs 2013;26(2):119-23.

[12]   Samer CF, Lorenzini KI, Rollason V, Daali Y, Desmeules JA. Applications of CYP450 testing in the clinical setting. *Mol. Diagn. Ther.* 2013;17(3):165-84.

[13]   Chen CH, Wang SC, Tsou HH, et al. Genetic polymorphisms in CYP3A4 are associated with withdrawal symptoms and adverse reactions in methadone maintenance patients. *Pharmacogenomics* 2011;12(10):1397-406.

[14]   Porcelli S, Fabbri C, Spina E, Serretti A, De Ronchi D. Genetic polymorphisms of cytochrome P450 enzymes and antidepressant metabolism. *Expert Opin. Drug Metab. Toxicol.* 2011;7(9):1101-15.

[15]   Soderberg MM, Haslemo T, Molden E, Dahl ML. Influence of CYP1A1/CYP1A2 and AHR polymorphisms on systemic olanzapine exposure. *Pharmacogenet Genomics* 2013;23(5):279-85.

[16]   Wang Z, Wang S, Huang M, Hu H, Yu L, Zeng S. Characterizing the Effect of Cytochrome P450 (CYP) 2C8, CYP2C9, and CYP2D6 Genetic Polymorphisms on Stereoselective N-demethylation of Fluoxetine. *Chirality* 2014;26(3):166-73.

[17]   Armahizer MJ, Seybert AL, Smithburger PL, Kane-Gill SL. Drug-drug interactions contributing to QT prolongation in cardiac intensive care units. *J. Crit. Care* 2013;28(3):243-9.

[18]   Azuma J, Hasunuma T, Kubo M, et al. The relationship between clinical pharmacokinetics of aripiprazole and CYP2D6 genetic polymorphism:

effects of CYP enzyme inhibition by coadministration of paroxetine or fluvoxamine. *Eur. J. Clin. Pharmacol.* 2012;68(1):29-37.

[19] Conley RR, Kelly DL. Drug-drug interactions associated with second-generation antipsychotics: considerations for clinicians and patients. *Psychopharmacol. Bull.* 2007;40(1):77-97.

[20] DeVane CL, Donovan JL, Liston HL, et al. Comparative CYP3A4 inhibitory effects of venlafaxine, fluoxetine, sertraline, and nefazodone in healthy volunteers. *J. Clin. Psychopharmacol.* 2004;24(1):4-10.

[21] DeVane CL, Nemeroff CB. An evaluation of risperidone drug interactions. *J. Clin. Psychopharmacol.* 2001;21(4):408-16.

[22] Greenblatt DJ, von Moltke LL, Harmatz JS, Shader RI. Human cytochromes and some newer antidepressants: kinetics, metabolism, and drug interactions. J. Clin. Psychopharmacol. 1999;19(5 Suppl 1):23S-35S.

[23] Gronlund J, Saari TI, Hagelberg NM, Neuvonen PJ, Laine K, Olkkola KT. Effect of inhibition of cytochrome P450 enzymes 2D6 and 3A4 on the pharmacokinetics of intravenous oxycodone: a randomized, three-phase, crossover, placebo-controlled study. *Clin. Drug Investig.* 2011;31(3):143-53.

[24] Miguel C, Albuquerque E. Drug interaction in psycho-oncology: antidepressants and antineoplastics. *Pharmacology* 2011;88(5-6):333-9.

[25] Preskorn SH, Shah R, Neff M, Golbeck AL, Choi J. The potential for clinically significant drug-drug interactions involving the CYP 2D6 system: effects with fluoxetine and paroxetine versus sertraline. *J. Psychiatr. Pract.* 2007;13(1):5-12.

[26] Prior TI, Baker GB. Interactions between the cytochrome P450 system and the second-generation antipsychotics. *J. Psychiatry Neurosci.* 2003;28(2):99-112.

[27] Sawada Y, Satoh H. [Drug interactions of psychotropic drugs]. *Nihon Rinsho* 2012;70(1):27-41.

[28] Spina E, de Leon J. Metabolic drug interactions with newer antipsychotics: a comparative review. *Basic Clin. Pharmacol. Toxicol.* 2007;100(1):4-22.

[29] Schosser A, Kasper S. The role of pharmacogenetics in the treatment of depression and anxiety disorders. *Int. Clin. Psychopharmacol.* 2009;24(6):277-88.

[30] Loovers HM, van der Weide J. Implementation of CYP2D6 genotyping in psychiatry. *Expert Opin. Drug Metab. Toxicol.* 2009;5(9):1065-77.

[31] Witherspoon DJ, Wooding S, Rogers AR, et al. Genetic similarities within and between human populations. *Genetics* 2007;176(1):351-9.

[32] Witherspoon DJ, Marchani EE, Watkins WS, et al. Human population genetic structure and diversity inferred from polymorphic L1(LINE-1) and Alu insertions. *Hum. Hered.* 2006;62(1):30-46.

[33] Jorde LB, Wooding SP. Genetic variation, classification and 'race'. *Nat. Genet* 2004;36(11 Suppl):S28-33.

[34] Aklillu E, Persson I, Bertilsson L, Johansson I, Rodrigues F, Ingelman-Sundberg M. Frequent distribution of ultrarapid metabolizers of debrisoquine in an ethiopian population carrying duplicated and multiduplicated functional CYP2D6 alleles. *J. Pharmacol. Exp. Ther.* 1996;278(1):441-6.

[35] Merlin MD. Archaeological Evidence for the Tradition of Psychoactive Plant Use in the Old World. *Economic Botany* 2003;57(3):295–323.

[36] Stolberg VB. The use of coca: prehistory, history, and ethnography. *J. Ethn. Subst. Abuse* 2011;10(2):126-46.

[37] Wall CA, Catherine M Oldenkamp, MMSII, Cosima Swintak, MD. Safety and Efficacy Pharmacogenomics in Pediatric Psychopharmacology. Primary Psychiatry 2010; *Clinical Focus* (17):53-8.

[38] Hendset M, Hermann M, Lunde H, Refsum H, Molden E. Impact of the CYP2D6 genotype on steady-state serum concentrations of aripiprazole and dehydroaripiprazole. *Eur. J. Clin. Pharmacol.* 2007;63(12):1147-51.

[39] Hendset M, Molden E, Knape M, Hermann M. Serum concentrations of risperidone and aripiprazole in subgroups encoding CYP2D6 intermediate metabolizer phenotype. *Ther. Drug Monit.* 2014;36(1):80-5.

[40] Urichuk L, Prior TI, Dursun S, Baker G. Metabolism of atypical antipsychotics: involvement of cytochrome p450 enzymes and relevance for drug-drug interactions. *Curr. Drug Metab.* 2008;9(5):410-8.

[41] FDA. US Food and Drug Administration. http://wwwaccessdatafdagov/scripts/cder/drugsatfda/indexcfm.

[42] PharmGkb. http://www.pharmgkb.org/search/browseAlpha.action?browseKey=variantAnnotatedGenes.

[43] Swen JJ, Nijenhuis M, de Boer A, et al. Pharmacogenetics: from bench to byte--an update of guidelines. *Clin. Pharmacol. Ther.* 2011;89(5):662-73.

[44] Kubo M, Koue T, Maune H, Fukuda T, Azuma J. Pharmacokinetics of aripiprazole, a new antipsychotic, following oral dosing in healthy adult

Japanese volunteers: influence of CYP2D6 polymorphism. *Drug Metab. Pharmacokinet.* 2007;22(5):358-66.

[45] Kim SH, Ivanova O, Abbasi FA, Lamendola CA, Reaven GM, Glick ID. Metabolic impact of switching antipsychotic therapy to aripiprazole after weight gain: a pilot study. *J. Clin. Psychopharmacol.* 2007;27(4):365-8.

[46] Kubo M, Koue T, Inaba A, et al. Influence of itraconazole co-administration and CYP2D6 genotype on the pharmacokinetics of the new antipsychotic ARIPIPRAZOLE. *Drug Metab. Pharmacokinet.* 2005;20(1):55-64.

[47] Suzuki T, Mihara K, Nakamura A, et al. Effects of the CYP2D6*10 allele on the steady-state plasma concentrations of aripiprazole and its active metabolite, dehydroaripiprazole, in Japanese patients with schizophrenia. *Ther. Drug Monit.* 2011;33(1):21-4.

[48] Doude van Troostwijk LJ, Koopmans RP, Vermeulen HD, Guchelaar HJ. CYP1A2 activity is an important determinant of clozapine dosage in schizophrenic patients. *Eur. J. Pharm. Sci.* 2003;20(4-5):451-7.

[49] Jaquenoud Sirot E, Knezevic B, Morena GP, et al. ABCB1 and cytochrome P450 polymorphisms: clinical pharmacogenetics of clozapine. *J. Clin. Psychopharmacol.* 2009;29(4):319-26.

[50] Murray M. Role of CYP pharmacogenetics and drug-drug interactions in the efficacy and safety of atypical and other antipsychotic agents. *J. Pharm. Pharmacol.* 2006;58(7):871-85.

[51] Olesen OV, Linnet K. Contributions of five human cytochrome P450 isoforms to the N-demethylation of clozapine in vitro at low and high concentrations. *J. Clin. Pharmacol.* 2001;41(8):823-32.

[52] Palma BB, Silva ESM, Urban P, Rueff J, Kranendonk M. Functional characterization of eight human CYP1A2 variants: the role of cytochrome b5. *Pharmacogenet. Genomics* 2013;23(2):41-52.

[53] Shad MU. Clozapine toxicity: A discussion of pharmacokinetic factors. *Asian J. Psychiatr.* 2008;1(2):47-9.

[54] Spina E, Scordo MG, D'Arrigo C. Metabolic drug interactions with new psychotropic agents. *Fundam. Clin. Pharmacol.* 2003;17(5):517-38.

[55] Ito K, Kubota Y, Toda T, Suto S, Ikarashi N, Sugiyama K. Effect of hydrocotarnine on cytochrome P450 and P-glycoprotein. *Drug Metab. Pharmacokinet.* 2009;24(1):108-13.

[56] Crews KR, Gaedigk A, Dunnenberger HM, et al. Clinical Pharmacogenetics Implementation Consortium (CPIC) guidelines for codeine therapy in the context of cytochrome P450 2D6 (CYP2D6) genotype. *Clin. Pharmacol. Ther.* 2012;91(2):321-6.

[57] Thorn CF, Klein TE, Altman RB. Codeine and morphine pathway. *Pharmacogenet. Genomics* 2009;19(7):556-8.

[58] Johansson I, Ingelman-Sundberg M. Genetic polymorphism and toxicology--with emphasis on cytochrome p450. *Toxicol. Sci.* 2011;120(1):1-13.

[59] Madadi P, Ciszkowski C, Gaedigk A, et al. Genetic transmission of cytochrome P450 2D6 (CYP2D6) ultrarapid metabolism: implications for breastfeeding women taking codeine. *Curr. Drug Saf.* 2011;6(1):36-9.

[60] Rogers JF, Nafziger AN, Bertino JS, Jr. Pharmacogenetics affects dosing, efficacy, and toxicity of cytochrome P450-metabolized drugs. *Am. J. Med.* 2002;113(9):746-50.

[61] Wilcox RA, Owen H. Variable cytochrome P450 2D6 expression and metabolism of codeine and other opioid prodrugs: implications for the Australian anaesthetist. *Anaesth. Intensive Care* 2000;28(6):611-9.

[62] Berka K, Anzenbacherova E, Hendrychova T, et al. Binding of quinidine radically increases the stability and decreases the flexibility of the cytochrome P450 2D6 active site. *J. Inorg. Biochem.* 2012;110:46-50.

[63] VandenBrink BM, Foti RS, Rock DA, Wienkers LC, Wahlstrom JL. Prediction of CYP2D6 drug interactions from in vitro data: evidence for substrate-dependent inhibition. *Drug Metab. Dispos.* 2012;40(1):47-53.

[64] Ai C, Li Y, Wang Y, Chen Y, Yang L. Insight into the effects of chiral isomers quinidine and quinine on CYP2D6 inhibition. *Bioorg. Med. Chem. Lett.* 2009;19(3):803-6.

[65] Luk S, Atayee RS, Ma JD, Best BM. Urinary diazepam metabolite distribution in a chronic pain population. *J. Anal. Toxicol.* 2014; 38(3):135-42.

[66] Andresen H, Augustin C, Streichert T. Toxicogenetics--cytochrome P450 microarray analysis in forensic cases focusing on morphine/codeine and diazepam. *Int. J. Legal Med.* 2013;127(2):395-404.

[67] Mandrioli R, Mercolini L, Raggi MA. Benzodiazepine metabolism: an analytical perspective. *Curr. Drug Metab.* 2008;9(8):827-44.

[68] Riss J, Cloyd J, Gates J, Collins S. Benzodiazepines in epilepsy: pharmacology and pharmacokinetics. *Acta Neurol. Scand.* 2008;118(2): 69-86.

[69] Walzer M, Bekersky I, Blum RA, Tolbert D. Pharmacokinetic drug interactions between clobazam and drugs metabolized by cytochrome P450 isoenzymes. *Pharmacotherapy* 2012;32(4):340-53.

[70] MIMsOnline   C.   https://www.mimsonline.com.au.acs.hcn.com.au/ Search/Search.aspx.

[71] Singh MS, Francis PA, Michael M. Tamoxifen, cytochrome P450 genes and breast cancer clinical outcomes. *Breast* 2011;20(2):111-8.

[72] Spina E, Santoro V, D'Arrigo C. Clinically relevant pharmacokinetic drug interactions with second-generation antidepressants: an update. *Clin. Ther.* 2008;30(7):1206-27.

[73] Molden E, Garcia BH, Braathen P, Eggen AE. Co-prescription of cytochrome P450 2D6/3A4 inhibitor-substrate pairs in clinical practice. A retrospective analysis of data from Norwegian primary pharmacies. *Eur. J. Clin. Pharmacol.* 2005;61(2):119-25.

[74] Hemeryck A, Belpaire FM. Selective serotonin reuptake inhibitors and cytochrome P-450 mediated drug-drug interactions: an update. *Curr. Drug Metab.* 2002;3(1):13-37.

[75] Hiemke C, Hartter S. Pharmacokinetics of selective serotonin reuptake inhibitors. *Pharmacol. Ther.* 2000;85(1):11-28.

[76] Sproule BA, Naranjo CA, Brenmer KE, Hassan PC. Selective serotonin reuptake inhibitors and CNS drug interactions. A critical review of the evidence. *Clin. Pharmacokinet.* 1997;33(6):454-71.

[77] Seltzer B. Galantamine-ER for the treatment of mild-to-moderate Alzheimer's disease. *Clin. Interv. Aging* 2010;5:1-6.

[78] Mannheimer B, Wettermark B, Lundberg M, Pettersson H, von Bahr C, Eliasson E. Nationwide drug-dispensing data reveal important differences in adherence to drug label recommendations on CYP2D6-dependent drug interactions. *Br. J. Clin. Pharmacol.* 2010;69(4):411-7.

[79] Huang F, Fu Y. A review of clinical pharmacokinetics and pharmacodynamics of galantamine, a reversible acetylcholinesterase inhibitor for the treatment of Alzheimer's disease, in healthy subjects and patients. *Curr. Clin. Pharmacol.* 2010;5(2):115-24.

[80] Farlow MR. Clinical pharmacokinetics of galantamine. *Clin. Pharmacokinet.* 2003;42(15):1383-92.

[81] Tavassoli N, Sommet A, Lapeyre-Mestre M, Bagheri H, Montrastruc JL. Drug interactions with cholinesterase inhibitors: an analysis of the French pharmacovigilance database and a comparison of two national drug formularies (Vidal, British National Formulary). *Drug Saf.* 2007;30(11):1063-71.

[82] Huang F, Lasseter KC, Janssens L, Verhaeghe T, Lau H, Zhao Q. Pharmacokinetic and safety assessments of galantamine and risperidone

after the two drugs are administered alone and together. *J. Clin. Pharmacol.* 2002;42(12):1341-51.

[83] Potkin SG, Preskorn S, Hochfeld M, Meng X. A thorough QTc study of 3 doses of iloperidone including metabolic inhibition via CYP2D6 and/or CYP3A4 and a comparison to quetiapine and ziprasidone. *J. Clin. Psychopharmacol.* 2013;33(1):3-10.

[84] Citrome L. Iloperidone: a clinical overview. *J. Clin. Psychiatry* 2011;72 Suppl 1:19-23.

[85] Fijal BA, Stauffer VL, Kinon BJ, et al. Analysis of gene variants previously associated with iloperidone response in patients with schizophrenia who are treated with risperidone. *J. Clin. Psychiatry* 2012;73(3):367-71.

[86] Arif SA, Mitchell MM. Iloperidone: A new drug for the treatment of schizophrenia. *Am. J. Health Syst. Pharm.* 2011;68(4):301-8.

[87] Citrome L. Iloperidone: chemistry, pharmacodynamics, pharmacokinetics and metabolism, clinical efficacy, safety and tolerability, regulatory affairs, and an opinion. *Expert Opin. Drug Metab. Toxicol.* 2010;6(12):1551-64.

[88] Robertson P, Jr., Hellriegel ET. Clinical pharmacokinetic profile of modafinil. *Clin. Pharmacokinet.* 2003;42(2):123-37.

[89] Robertson P, DeCory HH, Madan A, Parkinson A. In vitro inhibition and induction of human hepatic cytochrome P450 enzymes by modafinil. *Drug Metab. Dispos.* 2000;28(6):664-71.

[90] Scott SA, Sangkuhl K, Shuldiner AR, et al. PharmGKB summary: very important pharmacogene information for cytochrome P450, family 2, subfamily C, polypeptide 19. *Pharmacogenet. Genomics* 2012;22(2): 159-65.

[91] Swen JJ, Wilting I, de Goede AL, et al. Pharmacogenetics: from bench to byte. *Clin. Pharmacol. Ther.* 2008;83(5):781-7.

[92] Wenzel-Seifert K, Wittmann M, Haen E. QTc prolongation by psychotropic drugs and the risk of Torsade de Pointes. *Dtsch Arztebl. Int.* 2011;108(41):687-93.

[93] Gongadze N, Kezeli T, Antelava N. Prolong QT interval and "torsades de pointes" associated with different group of drugs. *Georgian Med. News* 2007(153):45-9.

[94] Zhou SF. Polymorphism of human cytochrome P450 2D6 and its clinical significance: part II. *Clin. Pharmacokinet.* 2009;48(12):761-804.

[95] Sweet RA, Pollock BG, Mulsant BH, et al. Pharmacologic profile of perphenazine's metabolites. *J. Clin. Psychopharmacol.* 2000;20(2):181-7.

[96] Rogers HL, Bhattaram A, Zineh I, et al. CYP2D6 genotype information to guide pimozide treatment in adult and pediatric patients: basis for the U.S. Food and Drug Administration's new dosing recommendations. *J. Clin. Psychiatry* 2012;73(9):1187-90.

[97] Desta Z, Kerbusch T, Soukhova N, Richard E, Ko JW, Flockhart DA. Identification and characterization of human cytochrome P450 isoforms interacting with pimozide. *J. Pharmacol. Exp. Ther.* 1998;285(2):428-37.

[98] Suzuki Y, Fukui N, Tsuneyama N, et al. Effect of the cytochrome P450 2D6*10 allele on risperidone metabolism in Japanese psychiatric patients. *Hum. Psychopharmacol.* 2012;27(1):43-6.

[99] Zhou SF. Polymorphism of human cytochrome P450 2D6 and its clinical significance: Part I. *Clin. Pharmacokinet.* 2009;48(11):689-723.

[100] Mannheimer B, von Bahr C, Pettersson H, Eliasson E. Impact of multiple inhibitors or substrates of cytochrome P450 2D6 on plasma risperidone levels in patients on polypharmacy. *Ther. Drug Monit.* 2008;30(5):565-9.

[101] Fang J, Bourin M, Baker GB. Metabolism of risperidone to 9-hydroxyrisperidone by human cytochromes P450 2D6 and 3A4. *Naunyn Schmiedebergs Arch. Pharmacol.* 1999;359(2):147-51.

[102] Guay DR. Tetrabenazine, a monoamine-depleting drug used in the treatment of hyperkinetic movement disorders. *Am. J. Geriatr. Pharmacother.* 2010;8(4):331-73.

[103] Dorado P, Penas LEM, de la Rubia A, A LL. Relevance of CYP2D6 - 1584C>G polymorphism for thioridazine:mesoridazine plasma concentration ratio in psychiatric patients. *Pharmacogenomics* 2009;10(7):1083-9.

[104] Thanacoody RH, Daly AK, Reilly JG, Ferrier IN, Thomas SH. Factors affecting drug concentrations and QT interval during thioridazine therapy. *Clin. Pharmacol. Ther.* 2007;82(5):555-65.

[105] Wojcikowski J, Maurel P, Daniel WA. Characterization of human cytochrome p450 enzymes involved in the metabolism of the piperidine-type phenothiazine neuroleptic thioridazine. *Drug Metab. Dispos.* 2006;34(3):471-6.

[106] Nakagami T, Yasui-Furukori N, Saito M, et al. Thioridazine inhibits risperidone metabolism: a clinically relevant drug interaction. *J. Clin. Psychopharmacol.* 2005;25(1):89-91.

[107] Berecz R, de la Rubia A, Dorado P, Fernandez-Salguero P, Dahl ML, A LL. Thioridazine steady-state plasma concentrations are influenced by tobacco smoking and CYP2D6, but not by the CYP2C9 genotype. *Eur. J. Clin. Pharmacol.* 2003;59(1):45-50.

[108] A LL, Berecz R, de la Rubia A, Dorado P. QTc interval lengthening is related to CYP2D6 hydroxylation capacity and plasma concentration of thioridazine in patients. *J. Psychopharmacol.* 2002;16(4):361-4.

[109] A LL, Berecz R, de la Rubia A, Fernandez-Salguero P, Dorado P. Effect of thioridazine dosage on the debrisoquine hydroxylation phenotype in psychiatric patients with different CYP2D6 genotypes. *Ther. Drug Monit.* 2001;23(6):616-20.

[110] Allegaert K, Rochette A, Veyckemans F. Developmental pharmacology of tramadol during infancy: ontogeny, pharmacogenetics and elimination clearance. *Paediatr. Anaesth.* 2011;21(3):266-73.

[111] Allegaert K, van den Anker JN, de Hoon JN, et al. Covariates of tramadol disposition in the first months of life. *Br. J. Anaesth.* 2008; 100(4):525-32.

[112] Gan SH, Ismail R, Wan Adnan WA, Zulmi W. Impact of CYP2D6 genetic polymorphism on tramadol pharmacokinetics and pharmacodynamics. *Mol. Diagn. Ther.* 2007;11(3):171-81.

[113] Grond S, Sablotzki A. Clinical pharmacology of tramadol. *Clin. Pharmacokinet.* 2004;43(13):879-923.

[114] Levo A, Koski A, Ojanpera I, Vuori E, Sajantila A. Post-mortem SNP analysis of CYP2D6 gene reveals correlation between genotype and opioid drug (tramadol) metabolite ratios in blood. *Forensic Sci. Int.* 2003;135(1):9-15.

[115] Holliday SM, Benfield P. Venlafaxine. A review of its pharmacology and therapeutic potential in depression. *Drugs* 1995;49(2):280-94.

[116] Otton SV, Ball SE, Cheung SW, Inaba T, Rudolph RL, Sellers EM. Venlafaxine oxidation in vitro is catalysed by CYP2D6. *Br. J. Clin. Pharmacol.* 1996;41(2):149-56.

[117] Fogelman SM, Schmider J, Venkatakrishnan K, et al. O- and N-demethylation of venlafaxine in vitro by human liver microsomes and by microsomes from cDNA-transfected cells: effect of metabolic inhibitors and SSRI antidepressants. *Neuropsychopharmacology* 1999;20(5):480-90.

[118] Kingback M, Karlsson L, Zackrisson AL, et al. Influence of CYP2D6 genotype on the disposition of the enantiomers of venlafaxine and its major metabolites in postmortem femoral blood. *Forensic Sci. Int.* 2012;214(1-3):124-34.

[119] Klamerus KJ, Maloney K, Rudolph RL, Sisenwine SF, Jusko WJ, Chiang ST. Introduction of a composite parameter to the pharmacokinetics of venlafaxine and its active O-desmethyl metabolite. *J. Clin. Pharmacol.* 1992;32(8):716-24.

[120] Macaluso M, Preskorn SH. CYP 2D6 PM status and antidepressant response to nortriptyline and venlafaxine: is it more than just drug metabolism? *J. Clin. Psychopharmacol.* 2011;31(2):143-5.

[121] Nichols AI, Focht K, Jiang Q, Preskorn SH, Kane CP. Pharmacokinetics of venlafaxine extended release 75 mg and desvenlafaxine 50 mg in healthy CYP2D6 extensive and poor metabolizers: a randomized, open-label, two-period, parallel-group, crossover study. *Clin. Drug Investig.* 2011;31(3):155-67.

[122] Oganesian A, Shilling AD, Young-Sciame R, et al. Desvenlafaxine and venlafaxine exert minimal in vitro inhibition of human cytochrome P450 and P-glycoprotein activities. *Psychopharmacol. Bull.* 2009;42(2):47-63.

[123] Preskorn S, Patroneva A, Silman H, et al. Comparison of the pharmacokinetics of venlafaxine extended release and desvenlafaxine in extensive and poor cytochrome P450 2D6 metabolizers. *J. Clin. Psychopharmacol.* 2009;29(1):39-43.

[124] Preskorn SH. Understanding outliers on the usual dose-response curve: venlafaxine as a way to phenotype patients in terms of their CYP 2D6 status and why it matters. *J. Psychiatr. Pract.* 2010;16(1):46-9.

[125] Shams ME, Arneth B, Hiemke C, et al. CYP2D6 polymorphism and clinical effect of the antidepressant venlafaxine. *J. Clin. Pharm. Ther.* 2006;31(5):493-502.

[126] Veefkind AH, Haffmans PM, Hoencamp E. Venlafaxine serum levels and CYP2D6 genotype. *Ther. Drug Monit.* 2000;22(2):202-8.

[127] Whirl-Carrillo M, McDonagh EM, Hebert JM, et al. Pharmacogenomics knowledge for personalized medicine. *Clin. Pharmacol. Ther.* 2012; 92(4):414-7.

[128] Picotte JJ, Rosenthal DM, Rhode JM, Cruzan MB. Plastic responses to temporal variation in moisture availability: consequences for water use efficiency and plant performance. *Oecologia* 2007;153(4):821-32.

[129] Sangkuhl K, Klein TE, Altman RB. PharmGKB summary: citalopram pharmacokinetics pathway. *Pharmacogenet Genomics* 2011;21(11):769-72.

[130] von Moltke LL, Greenblatt DJ, Grassi JM, et al. Citalopram and desmethylcitalopram in vitro: human cytochromes mediating transformation, and cytochrome inhibitory effects. *Biol. Psychiatry* 1999;46(6):839-49.

[131] Rochat B, Amey M, Gillet M, Meyer UA, Baumann P. Identification of three cytochrome P450 isozymes involved in N-demethylation of citalopram enantiomers in human liver microsomes. *Pharmacogenetics* 1997;7(1):1-10.

[132] Olesen OV, Linnet K. Studies on the stereoselective metabolism of citalopram by human liver microsomes and cDNA-expressed cytochrome P450 enzymes. *Pharmacology* 1999;59(6):298-309.

[133] Matsui E, Hoshino M, Matsui A, Okahira A. Simultaneous determination of citalopram and its metabolites by high-performance liquid chromatography with column switching and fluorescence detection by direct plasma injection. *J. Chromatogr. B. Biomed. Appl.* 1995;668(2):299-307.

[134] Sidhu J, Priskorn M, Poulsen M, Segonzac A, Grollier G, Larsen F. Steady-state pharmacokinetics of the enantiomers of citalopram and its metabolites in humans. *Chirality* 1997;9(7):686-92.

[135] Rocha A, Marques MP, Coelho EB, Lanchote VL. Enantioselective analysis of citalopram and demethylcitalopram in human and rat plasma by chiral LC-MS/MS: application to pharmacokinetics. *Chirality* 2007;19(10):793-801.

[136] Bondolfi G, Chautems C, Rochat B, Bertschy G, Baumann P. Non-response to citalopram in depressive patients: pharmacokinetic and clinical consequences of a fluvoxamine augmentation. *Psychopharmacology* (Berl) 1996;128(4):421-5.

[137] Berzas-Nevado JJ, Villasenor-Llerena MJ, Guiberteau-Cabanillas C, Rodriguez-Robledo V. Enantiomeric screening of racemic citalopram and metabolites in human urine by entangled polymer solution capillary electrophoresis: an innovatory robustness/ruggedness study. *Electrophoresis* 2006;27(4):905-17.

[138] Rochat B, Kosel M, Boss G, Testa B, Gillet M, Baumann P. Stereoselective biotransformation of the selective serotonin reuptake inhibitor citalopram and its demethylated metabolites by monoamine oxidases in human liver. *Biochem. Pharmacol.* 1998;56(1):15-23.

[139] Ohno S, Kawana K, Nakajin S. Contribution of UDP-glucuronosyltransferase 1A1 and 1A8 to morphine-6-glucuronidation and its kinetic properties. *Drug Metab. Dispos.* 2008;36(4):688-94.

[140] Milne RW, Nation RL, Somogyi AA. The disposition of morphine and its 3- and 6-glucuronide metabolites in humans and animals, and the importance of the metabolites to the pharmacological effects of morphine. *Drug Metab. Rev.* 1996;28(3):345-472.

[141] Caraco Y, Tateishi T, Guengerich FP, Wood AJ. Microsomal codeine N-demethylation: cosegregation with cytochrome P4503A4 activity. *Drug Metab. Dispos.* 1996;24(7):761-4.

[142] Chen ZR, Somogyi AA, Reynolds G, Bochner F. Disposition and metabolism of codeine after single and chronic doses in one poor and seven extensive metabolisers. *Br. J. Clin. Pharmacol.* 1991;31(4):381-90.

[143] Yue QY, Hasselstrom J, Svensson JO, Sawe J. Pharmacokinetics of codeine and its metabolites in Caucasian healthy volunteers: comparisons between extensive and poor hydroxylators of debrisoquine. *Br. J. Clin. Pharmacol.* 1991;31(6):635-42.

[144] Madadi P, Koren G. Pharmacogenetic insights into codeine analgesia: implications to pediatric codeine use. *Pharmacogenomics* 2008;9(9): 1267-84.

[145] Sangkuhl Katrin SJC, Turpeinen Miia, Altman Russ B, Klein Teri E. PharmGKB summary: venlafaxine pathway. *Pharmacogenet. Genomics* 2013.

[146] Eap CB, Bertel-Laubscher R, Zullino D, Amey M, Baumann P. Marked increase of venlafaxine enantiomer concentrations as a consequence of metabolic interactions: a case report. *Pharmacopsychiatry* 2000; 33(3):112-5.

INDEX

H

M